An Urban Geography of England and Wales in the Nineteenth Century

Harold Carter

Emeritus Gregynog Professor of Human Geography,
University College of Wales, Aberystwyth

and

C. Roy Lewis

Head of Department of Geography, Institute of Earth
Studies, University College of Wales, Aberystwyth

Edward Arnold
A division of Hodder & Stoughton
LONDON NEW YORK MELBOURNE AUCKLAND

© 1990 Harold Carter and C. Roy Lewis

First published in Great Britain 1990

Distributed in the USA by Routledge, Chapman and Hall, Inc.
29 West 35th Street, New York, NY 10001

British Library Cataloguing in Publication Data
Carter, Harold
  An Urban Geography of England and Wales in the
  Nineteenth Century
  1. Great Britain. Cities, history
  I. Title   II. Lewis, Roy C.
  307.7640941

  ISBN 0-7131-6549-9

22112153

Typeset in Linotron Sabon by Rowland Phototypesetting Limited,
Bury St Edmunds, Suffolk. Printed in Great Britain for
Edward Arnold, a division of Hodder and Stoughton Limited,
Mill Road, Dunton Green, Sevenoaks, Kent TN13 2YA by
St Edmundsbury Press Limited, Bury St Edmunds, Suffolk. Bound by
W. H. Ware & Son, Clevedon, Avon.

Contents

Acknowledgements

The authors and publishers are most grateful to the technical and secretarial staff in the Department of Geography, University College of Wales, Aberystwyth, for their cartographic, photographic and word-processing services – especially Miss G. Hamer, Mr D. Griffiths, Mr G. Hughes, Mr M. Rook, Mr I. Gulley and Mr M. G. Jones – and to Mrs Mari Carter who typed sections of the book. Dr Sandra Wheatley kindly provided Plate 11.1 and advice on various sources. The cover drawing, *Merthyr from Thomas Town*, printed by Rock and Co., London, in 1865, was supplied by Merthyr Reference Library.

1

Introduction

To embark on a study of nineteenth-century urbanism, even though it be limited to England and Wales, is not, to continue the metaphor, to be

> . . . the first that ever burst
> Into that silent sea.

Rather it is to enter an ocean much navigated by ships of all shapes and sizes, though hardly sailing in constant directions on established trade routes. As the concern of historians and social scientists with the last century has developed, a great range of different interests and approaches has been generated. The standard two-volume work on *The Victorian City* edited by Dyos and Wolff (1982) extends across many academic disciplines. It includes, and they are quoted only for exemplification, studies of the urban element in nineteenth-century literature and of the informal 'literature of the streets', as well as contributions on such diverse themes as public health and religion, perhaps echoing Victorian concern with health in body and mind. Again, Anthony Sutcliffe in the book which he edited on *The Metropolis 1890–1940* (1984), a period which covers the end years of the century which this book intends to survey, included chapters on literature, music, the visual arts and the cinema.

This great diversity of material, this intellectual eclecticism, must surely be welcomed as an encouraging sign of interdisciplinary and multidisciplinary study, although the latter still preserving traditional subject areas, albeit in association, is much the more common. However, that same diversity must be taken as an indication that urban history has failed to find the unity it once appeared to seek as a distinctive, systematic branch of the general field of economic and social history. Not only in content but also in method that unity has been elusive, even chimerical. The earliest form of a proposed basis occurred in the adoption of quantification in what was boldly proclaimed as 'the new urban history' (Thernstrom and Sennett, 1969), but that rapidly gave way to disenchantment even before it was fully established. Retreat, if that be the proper interpretation, followed to Weberian studies of the urban community and to behavioural interpretations of the roles of interest groups and individuals. The alternative has been the rapid efflorescence of structuralist, specifically Marxist, demonstrations of how the inbuilt contradictions of the capitalist system have become manifest in its major creation, the industrial metropolis. These debates on interpretation have themselves become the progenitor of a prolix and lively urban academic literature. Works which marked the untimely death of H. J. Dyos (Cannadine and Reader, 1982; Fraser and Sutcliffe, 1983) and the many reviews of them, especially generated methodological debate to such an extent that there is but little purpose in retreading familiar ground in this introduction by adding to it. But in light of the plethora of approaches, methods and theoretical stances it is essential to present as clearly and briefly as possible the bases and assumptions against which this book has been written.

The approach here adopted, to be both blunt and unfashionable, is neither interdisciplinary nor multidisciplinary. That in no way means that the literature of other disciplines can be disregarded or treated as anything but an essential prerequisite. Hopefully it does not suggest that the book will be of no interest to non-geographers. What it does stress is that it is written from the integrity of a defined and particular viewpoint, that of the urban geographer.

The last sentence of the last paragraph can, however, be considered to beg the question, but the central implication is the pre-eminence of the spatial theme; to argue for anything else is to argue geography out of existence. But that simple statement seemingly stripped of the sophistications of methodological debate, by itself fails to provide a clear basis or a simple platform from which to proceed. The direct implication is that geography is essentially spatial analysis where inferences are derived from and processes identified by the examination of mapped patterns. The drawbacks, indeed the inherent intellectual dangers, arising from such an approach are widely recognized. The reverse procedure seems at first to be possibly more tenable, that is to seek out and examine those processes, themselves derived from the larger operation of economy and society which create urban patterns. The reference point of departure then becomes not the spatial character of the independent city but those broad socio-economic forces which make it what it is, and dependent on them as it is. Inevitably within the form of presentation in this book such an approach will in part become employed, but unless an absolutely single-minded reductionism, essentially Marxist, is employed, the search for process must be unbounded and the selection of any one set of processes must lack proper justification. This is the familiar empirical-inductive theoretical-deductive impasse. Induction without theory must lack form and justification, deduction without an empirical basis can only be abstract and unrelated to the real world. But the foundation of this book is not a theoretical stance but a pre-emptive concern with what historians like to call 'shapes on the ground', but which is better broadened into spatial analysis. The end is to examine the distribution of cities and distributions within cities, to identify the evolving spatial characteristic of British cities during a well-defined and formative period of time and to offer explanation by reference to whatever theories or processes suggest themselves as relevant and seem to be pertinent. This makes no claim for objectivity or neutrality, or for a value-free stance. But the eclecticism is not one of discipline for that is explicitly geographical, but rather of interpretation and explanation. In brief, the aim is to examine the spatial nature and structure of nineteenth-century cities rather than to establish and demonstrate the overarching applicability of a single theoretical system.

This viewpoint means, in turn, an out-of-hand rejection of some entity to which the epithet 'the Victorian city', carrying the implication of a uniform identity, can be applied. The starting point is the large number of towns, of different sizes and functional characters and with different rates and peaks of growth. The end point is the identification of generalizations which hold good and the establishment of the processes which were in operation.

One further point must be made here. One of the problems which a book of this sort must face is the delimitation of area. The widest spread to include Europe and the USA is not undertaken since patterns of land ownership and the nature of administrative intervention were so different. Britain in the nineteenth century meant the whole of the British islands. But although there were common elements it is difficult to treat even those diverse elements as one. The book, therefore, has been limited primarily to England and Wales. However, some references to Ireland and Scotland have been introduced where they best illustrate particular points.

The structure of the book is derived directly from its intent. This brief introduction is followed by a first section which considers the data available and on which examination can be based. Any analysis can only be as effective as the data permit and hence the need to begin from that point. Given that the period covered ranges from 1750 to 1914 even the census provides very variable information. The first half-century, 1750 to 1800, is bereft even of census data, whilst in the next half-century, although there are decennial censuses which provide good general information, they are inadequate for comprehensive urban studies. From 1851 to 1881 access to the census enumerators' books (Mills and Pearce, 1989) provides a rich filling to a sandwich, for after 1881, owing to the demands of confidentiality, the books are not yet available and other sources for detailed information have to be sought. Even more difficult is the acquisition of material of a behavioural or decision-making character. Contemporary survey, making use of questionnaires, can elicit the way the individual perceives the city environment and the way in which he or she operates in it, but such an approach is impossible for the city of the past. A range of secondary material can be used, however, derived from newspapers, novels and private accounts and diaries. But such information is self-selecting and constitutes a biased sample with no guarantee of being representative of any larger group. There is, therefore, a constant constraint on what can be achieved exercised by the data which can be recalled, and it is critical that this be understood at the outset.

The second section considers urbanism on the broadest scale during the period by reviewing the growth of the urban population and the development of the city system. It constitutes a study of urbanization. Three aspects of change need mention. The first two are more directly technological. The rapidly growing ability to move easily over increasing distances within the city brought about by tramcar and railway resulted in the progressive diminution in the friction of distance. This meant that the possibilities for the physical extension of the city were greatly increased and the restrictions which had constrained the size of the pre-industrial city were released. Agglomerations, or as Patrick Geddes called them in 1917, conurbations, of a completely new size were created. At the same time new technology engendered new locational demands so that not only was there a revolution in the size of the city but also in the patterning, for those demands generated new settlements. All this inevitably had major impact upon the structuring of the city system.

The third aspect of change is much wider ranging. All those features associated with the increasing numbers and size of cities can be packaged together under the concept of 'modernization'. But from this springboard the broadest of approaches develops where the urbanization of nineteenth-century Britain is seen as an aspect of the development of the world system, of the evolution of the world space economy. This at once reintroduces the question posed earlier as to where enquiry into the geographical character of urbanism should begin, with the narration of spatial fact or some universalist quest for a first cause and an over-arching theory. In his book *Industrialization and economic history: theses and conjectures* (1970), Jonathan Hughes writes, 'To understand anything . . . one needs to sort out the relevant antecedents. The presumed links in antecedent actions and the decisions of the historian regarding where he can 'cut into' these causally connected prior events and still produce a believable historical analysis is the colligation problem' (Hughes, 1970, 30). But this colligation problem is further exacerbated when the entry point is not simply related to a series of events but much more fundamentally to a process of evolution. Thus, in introducing the major changes which characterize urbanization between 1750 and 1914 one is involved, at the widest pitch, with the sorts of studies presented by Wallerstein (1979, 1980) and Braudel (1982, 1984) of the modern world system or, in a more direct but similar manner, with the interpretation presented in outline by David Harvey in *Social justice and the city* (1973) where, significantly, it is not considered irrelevant in the discussion of that issue to include a section on the origins of urbanism some five thousand years before the present when pervasive injustice is the central theme.

Since the book itself will not be organized about these broader interpretations, it is worth while in this introduction to digress and follow Harvey's argument. He begins with Karl Polyani's (1968) identification of three modes of economic integration, reciprocity, redistribution and market exchange which he equates in general terms with Fried's (1967) three forms of social organization, egalitarian, rank and stratified. Under reciprocity and egalitarianism towns are absent but redistribution of necessity involves flows of goods which converge upon and diverge from organizing centres. An urban hierarchy is therefore an integral part of such a system and with its operation a ranked society is associated. But with the development of market exchange fundamental modifications ensued.

> World empires [Rome was an example] were basically redistributive in economic form. No doubt they bred clusters of merchants who engaged in economic exchange (primarily long-distance trade) but such clusters, however large, were a minor part of the total economy and not fundamentally determinative of its fate. Such long-distance trade tended to be ... 'administered trade' and not market trade, utilizing 'ports of trade'. It was only with the emergence of the modern world economy in sixteenth-century Europe that we saw the full development and economic prominence of market trade. This was the system called capitalism. Capitalism and a world economy (that is, a single division of labour but multiple polities and cultures) are obverse sides of the same coin. One does not cause the other. We are really defining the same characteristics of the same coin (Wallerstein, 1979, 6).

But as Harvey contends (1973, 211), under capitalism the market is not concerned with commodities being sold to buy commodities but with money which is used to purchase commodities which are then resold, often in a transformed state, to procure money. Financial control is associated with the appropriation of the means of production creating class divisions which are the basis of a stratified social system. Following such a theme, therefore, nineteenth-century urbanization needs to be interpreted as part of the growth of the world economy and the capitalist system, just as do core–periphery regional contrasts, for they are both products of the same structuring processes.

Three direct ways in which these processes impacted upon urbanism can be discussed in exemplification. Harvey notes that 'merchant capitalism was not ... resistant to industrial capitalism ... Yet the urban centres were still, by and large, dominated by the rank society and ... manufacturing was regulated and controlled' (Harvey, 1973, 259). Because of such a circumstance industry was forced away into new 'rural' locations and new industrial towns appeared. It is not a wholly convincing case since there was a considerable degree of follow-on from mercantile capitalism and an argument can be deployed for stability, certainly in the higher reaches of the urban hierarchy, a theme which subsequent chapters will have to take up. But, accepting it for the moment, it provides the basic explanation for the new towns which so markedly extended the rank-array in the nineteenth century. This also provides the key, therefore, to changes in that rank-array, for the old hierarchical ordering was derived from merchant capitalism. It was disrupted and a new, greatly complex, industrially based hierarchy was initiated.

The second example is derived from the essential financial basis. It is one that has been urged by Vance (1971). In a rank society, land is held as the basis of that rank but in a capitalist system it becomes a commodity to be bought and sold, on which to speculate. Profitability becomes the key and a completely new basis for determining the ownership and use of land is created. All these processes of land development and building speculation which later sections of this book will need to discuss are generated by the nature of the capitalist economic system.

The third example is somewhat similar. As a society structured by rank gives way to one stratified by class so that stratification becomes expressed spatially by the segregation of residential areas and the eventual move to single-class social areas. Harvey writes that this was 'very important to the self-respect of people, but irrelevant to the basic economic structure of society' (Harvey, 1973, 281). Here then is the crux of the issue of spatial patterning. Effectively Harvey is maintaining that the critical cause is 'the basic economic structure' and that the distributed features are superstructural matters of relatively little importance. But so would be the time-related events of history and the emotions and character interrelations of literature, all thus reduced to a single, universal and Marxist explicand. But more relevant to the present volume, one would have to turn to the whole gamut of the modernization process, of the initiation and spread of the world capitalist system as the fundamental basis for the interpretation of the nineteenth-century city – as, indeed, of everything else related to that century. Such universalism is not the purpose of this book, which is at once more limited and more empirical and it is pertinent to explain this at the outset.

This diversion has thus returned to the basic theme of this chapter, the organization of a study of towns in nineteenth-century Britain from a geographical standpoint. The starting point is a clear focus on location and the organization of space. Having dealt therefore with the growth of towns, the location of towns and the structure of the city system, it is possible to proceed to an examination of internal organization. The immediately new element introduced by industrial growth was, self-evidently, the industrial region or tract. Dispersed and sporadic craft industries were replaced by extensive areas dominated by factories, ship-yards or mine-heads. A new and extensive demander of land was introduced into the city and its consideration forms a point of entry into all those uses and activities demanding space in towns. To articulate and consider these at this point would be to anticipate the second section of the book but characterizing them all, whatever their nature, were the two associated principles, both of which applied to industrial land, those of increasing specialization and segregation, so that the city became a mosaic of land-uses. But this implies that the sole force in the creation of that mosaic was that of demands for space competing with each other on a basis of the evaluation of price and convenience. However, demand was certainly not the sole arbiter for the controls of the supply of land and finance for development were as germane and these too have to be examined.

It has been implied that the period which is being considered was one where uninhibited market forces, demand and supply, operated without constraint to create the form of cities. Such of course was not the case, for throughout the nineteenth century there was constant mediation from government which was mainly expressed through concern with issues of public health. But once more, in interpretation, there are broader issues. 'From the materialistic point of view, state interventions (or the process of the political apparatus of bourgeois society) are essentially determined by the crisis-laden character of capitalist society and by related class confrontations' (Hirsch, 1981, 593). Or in other words, which state the same case, 'Beneath the appearance of social control over the evolution of the urban system lies the inexorable dynamic of a complex of land-contingent events that is essentially out of control. The capitalist state is thus caught up in a constantly escalating spiral of urban interventions' (Dear and Scott, 1981, 15). Again the approach in this volume is somewhat more modest than these far-reaching interpretations and the latter part of the book attempts to set out the way in which legislation in part determined the internal character of British cities with a progressively greater impact during the nineteenth century.

The empirical intentions which have been declared in this introduction should not give the impression that description is the main end. A study of 'state intervention in the nineteenth century cities' (Buck, 1981, 504–5) argues for three types of history, 'depending on the priority that is placed on theory, and the sort of relation established between theory and empirical data' (Buck, 1981, 504). These are called descriptive, explanatory and theoretical history. It is evident that this volume makes little claim to deal with the last, but it certainly intends to be explanatory.

At times when social scientists are constantly urged that their work should be 'relevant' and 'policy orientated', a study of nineteenth-century cities may be considered academically and intellectually indulgent. Such a view is grossly in error. At the most elementary level the city structures which the present has to handle were the creations of a great age of city growth and need to be understood as such in order to be effectively dealt with at the present. Furthermore, insight into how the urban processes operated in an age of industrial revolution can be of the greatest value for, as in the late nineteenth century, a new industrial revolution is transforming the urban structure of Britain.

2

The data: sources for the investigation of cities in the nineteenth century

The aim of this chapter is to examine the main sources of information which may be employed in geographical studies of towns in the nineteenth century. It runs the obvious risks of giving too brief an assessment of those sources which have been the subjects of detailed appraisals elsewhere, and of being too selective, but it has the two attractions of focusing attention on this very important matter and of obviating the need to keep returning to questions about data in the subsequent chapters. It is hoped that the discussion here will serve as a general background to data sources, one which can be built on by consulting some of the specific references which are cited in it.

At the outset, it can be suggested that there is a need for two broad sets of data, first, those data which facilitate the reconstruction of spatial patterns, and second, data which deal with the processes which shaped those patterns. In the first, the basic requirement is precise locational information on people, businesses and buildings; in the second, it is information on the decisions which lay behind those patterns, the decisions of householders, shopkeepers, manufacturers, landowners, builders, and so on. For the former, there are some standard sources which are fairly readily available and which have been used with great success. Discussion of these will take up most of this chapter. For the latter, there are no ubiquitous sources which stand out. Information on the decisions which were made, about moving house, starting a business, or building property, for example, must be pieced together by the painstaking research of extant personal papers. Such sources are fraught with difficulties – for instance, they are fragmentary for any one town, they are usually biased to particular social groups, and they are sometimes anecdotal – but, nonetheless, they are the most likely to yield useful information. Having made this distinction in data sources, it is proposed to organize the chapter around it by looking first, and at most length, at data on spatial patterns, and second, at sources of information on the decision-makers.

Analysis of spatial patterns

Census enumerators' books

For the study of the socio-economic mosaic the principal sources of information are the enumerators' books of the British decennial censuses which contain information on the composition of households. Setting aside the first four censuses of the

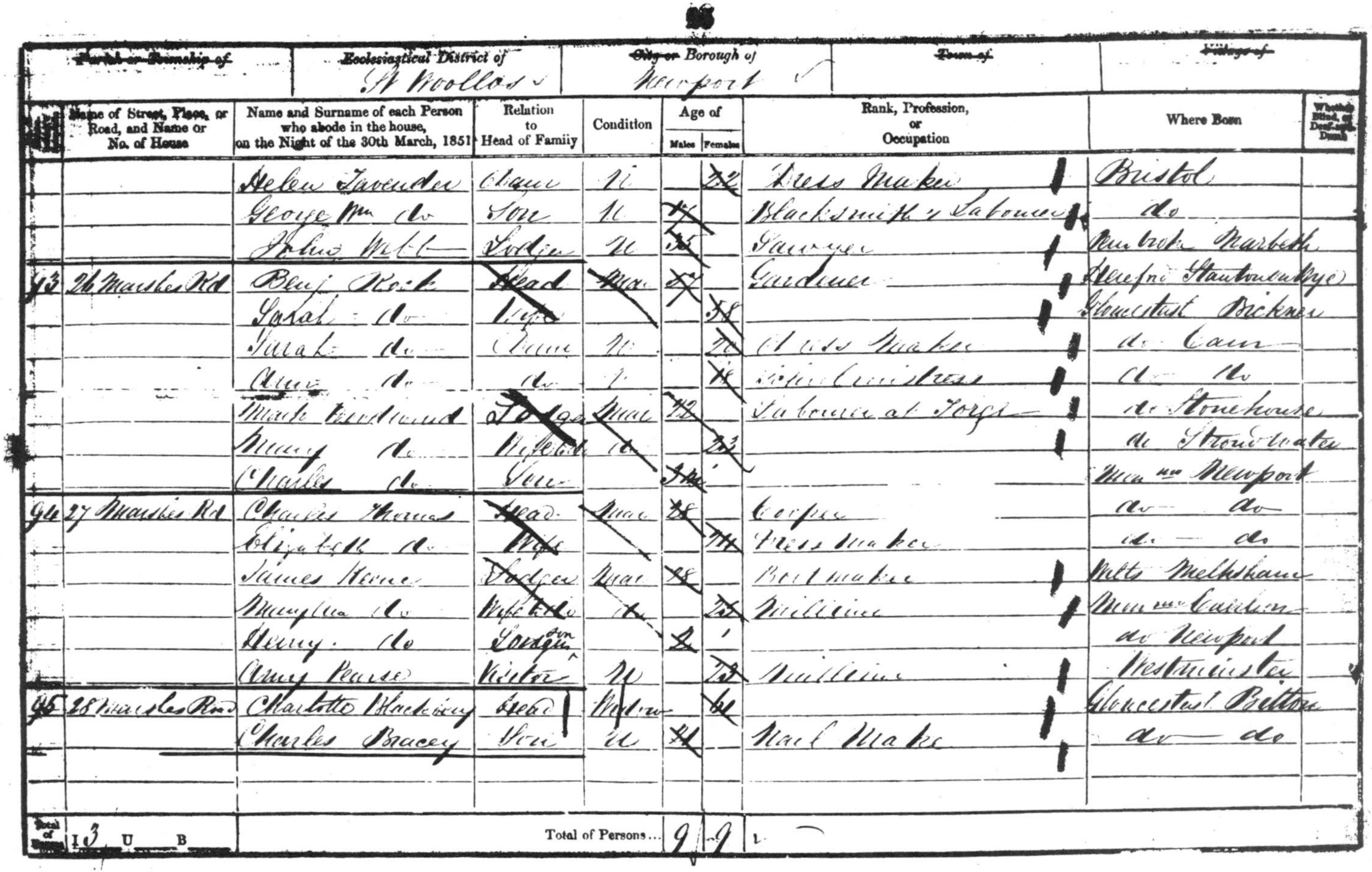

	Parish or Township of	Ecclesiastical District of St Woollos	City or Borough of Newport	Town of	Village of				
No. of Householder's Schedule	Name of Street, Place, or Road, and Name or No. of House	Name and Surname of each Person who abode in the house, on the Night of the 30th March, 1851	Relation to Head of Family	Condition	Age of Males	Age of Females	Rank, Profession, or Occupation	Where Born	Whether Blind, or Deaf-and-Dumb
		Helen Lavender	Dau	U		22	Dress Maker	Bristol	
		George Wm do	Son	U	2[illegible]		Blacksmith & Labourer	do	
		John Webb	Lodger	U	35		Sawyer	Pembroke Narbeth	
93	26 Marshes Rd	Benj Rook	Head	Mar	57		Gardener	Hereford Staunton on Wye	
		Sarah do	Wife			58		Gloucestershire Bickner	
		Sarah do	Dau	U		20	Dress Maker	do Cam	
		Amy do	do	U		18	School Mistress	do do	
		Mark Woodward	Lodger	Mar	22		Labourer at Forge	do Stonehouse	
		Mary do	Wife	do		23		do Stroudwater	
		Charles do	Son		3 m			Mon Newport	
94	27 Marshes Rd	Charles Thomas	Head	Mar	28		Cooper	do do	
		Elizabeth do	Wife			24	Dress Maker	do do	
		James Horne	Lodger	Mar	28		Boot maker	Wilts Melksham	
		Mary Ann do	Wife	do		22	Milliner	Mon Caerleon	
		Henry do	Lodger's son		2			do Newport	
		Amy Pearse	Visitor	U		23	Milliner	Westminster	
95	28 Marshes Road	Charlotte Blackwing	Head	Widow		61		Gloucestershire Bretton	
		Charles Bracey	Son	U	21		Nail Maker	do do	
Total of Houses 3 U B				Total of Persons...	9	9			

Figure 2.1 Sample page from the enumerators' books for Newport, Gwent, 1851 census. (Gwent County Record Office)

century, which were rather rudimentary in scope and methods of collection, a great deal of useful data is contained in the extant books for 1841, 1851, 1861, 1871 and 1881 (the last date being determined by the 100-year confidentiality rule barring access to information on individuals). Over recent years this material has been so widely discussed in published studies (for example, Lawton, 1978, and Mills, 1982) that it requires little attention here, except perhaps to point out its salient features and mention some of its drawbacks. This can be done by referring to Figure 2.1 which shows the basic layout which was adopted for enumerators' books from 1851 onwards. Along the top and in the two columns down the left-hand side of the page is locational information, specifying the town, parish, street and house. Then in sequence across the page are entered the details of all residents, including name, relation to the head of household, marital status, sex and age, occupation, place of birth and disabilities. Thus, in theory, from these details it is possible to reconstruct the composition of all households at the census dates.

However, in practice a number of difficulties arise. At the simplest level the legibility of handwriting can be a stumbling block. More fundamental, though, are problems of definition and interpretation which arise in the abstraction of data. For instance, there is the problem of identifying separate households in multi-occupied houses. The census enumerators employed various devices to differentiate between the end of a household and the end of a house, such as lines of various length ruled across the page at the divisions, but inconsistencies were quite common. Similarly, in recording occupations and birthplaces some enumerators were less punctilious than others. Then again, the accuracy of the data themselves has been called into question. For instance, Anderson in his study of Preston, pointed to some inaccuracies over age and birthplace when he compared the entries in successive censuses. Of 475 inhabitants he traced in both 1851 and 1861, 47 per cent had an age in 1861 more or less than 10 years older than in 1851. However, only 4 per cent were more than 2 years out and under 1 per cent more than 5 years in error. In terms of birthplace, at least 14 per cent of the 475 persons had a discrepancy between the two years; some were of no great importance, but in half of the cases it had a bearing on the numbers classified as migrants or non-migrants (Anderson, 1972, 75). Nonetheless, the enumerators' books are the most comprehensive source of data and have figured prominently in geographical studies.

The strength of the mid-century censuses is that they facilitate the analysis of family structure, employment and social status, and migration (insofar as migration can be re-traced through the birthplaces of parents and children). Out of the variables in the enumerators' books, perhaps most emphasis has been placed on occupation, not only as a measure of the employment structure of communities, but, more importantly, as an indicator of the social class of residents. As Armstrong has noted, occupation may be used to classify individuals 'according to two main principles, (a) by *industrial grouping* – so that we can trace the *economic* contours of a society and the bases on which these rest, and (b) by *social ranking*' (Armstrong, 1972, 191). However, the allocation of individuals to these industrial groups and social classes is by no means a straightforward exercise, and it has generated an extensive literature in its own right. A useful discussion of the main problems, and of possible solutions, has been provided by Armstrong (1972), and given its importance, the salient points will be repeated here.

The first standardized industrial classification was made at the 1911 census in which workers were placed into one of twenty-two industrial groups. Various modifications have been made at subsequent censuses. However, since these modern classifications cannot be used to organize the data collected in earlier censuses, it is necessary 'to devise a scheme which will retain many of the characteristics of the modern industrial distribution and take us beyond the vague and formless "occupational classes or orders" of nineteenth-century censuses' (Armstrong, 1972, 228). Armstrong has suggested that, with minor changes, the scheme devised by Charles Booth in the 1880s, in preparation for his *Life and labour of the people of London*, is an acceptable classification. Booth identified 11 major industrial categories of agriculture, fishing, mining, building, manufacture, transport, dealing, industrial service, public service and professional, domestic service, and 'others' (which embraced property owners and those of independent means, and indefinite occupations). In turn, these were subdivided into industrial subgroups and occupations. Using the manuscript notes made by Booth when he drew up his categories, Armstrong has presented a revised classification comprising nine major industrial sectors, which are subdivided into 79 industrial subgroups made up of 346 occupations. Not only can this standardized scheme be applied to individuals in the enumerators' books, but 'a uniform basis is thus made available for comparing communities with one another at any particular census date' (Armstrong, 1972, 247).

Likewise for the determination of the social status of individuals, Armstrong has promoted a standardized grouping derived from an existing classification. Although there are a number of modern schemes of stratification he argued that 'only one general scheme of social classification will be found to fit the two desiderata, i.e. not too refined for the data, and carrying with it published lists of occupations for easy allocation and comparability of classification, namely the Registrar-General's social classification scheme. Here all occupations are subsumed under five broad categories' (Armstrong, 1972, 202). As early as 1911 the Registrar-General grouped occupations into different social grades on the basis of their general standing within the community. This rough and ready attempt was replaced by a much more robust fivefold classification in 1921 under which for the first time individuals could be assigned to their appropriate social stratum. Revisions were made in 1931 and 1951, but the same five broad classes were retained, viz.

Class   I   Professional etc., occupations
Class II   Intermediate occupations
Class III   Skilled occupations
Class IV   Partly skilled occupations
Class V   Unskilled occupations.

The whole process of allocating occupations to these classes was undertaken by the Registrar-General as summarized in *Classification of occupations, 1950* (HMSO, 1951) where thousands of occupations were coded for social class.

Although this is a twentieth-century scheme, it has been accepted by many as a satisfactory classification for nineteenth-century occupations. In his study of the social structure of York, 1841–51, Armstrong adopted the 1951 system to classify residents, but he suggested a few modifications:

1   all those described as dealers, merchants, shopkeepers, innkeepers and such commercial traders should be placed initially in class III and not class II;
2   all employers of twenty-five or more persons should be placed in class I, and those in class III or IV who employed one or more persons should be elevated to class II;
3   drivers of horse-drawn passenger conveyances should be put in class III and not class IV.

Having made these modifications Armstrong went on to demonstrate its applicability to his 1851 sample of 753 heads of household.

While the Registrar-General's classification has been criticized for its retrospective application to the nineteenth century and its tendency to produce a bunching in class III, it possesses the clear advantages of being workable, of having ready-made tabulations of coded occupations, of picking out the extremes in society, and of ensuring comparability between communities. There are other schemes which have tried to iron out some of the inadequacies but which are a little more elaborate over the definition of classes. Royle, for instance, while retaining a five-class scheme, has suggested two modifications (Royle, 1977). First, he considered that the 'partly skilled' class IV of the Registrar-General had no place in the nineteenth-century context and abandoned it. Instead he subdivided class III into two parts, a new class III of non-manual occupations, and a new class IV of skilled manual occupations. Second, he introduced the keeping of servants as a further qualification for classes I, II and III. This modified scheme is shown in Table 2.1. Cowlard has also been critical of the raw Armstrong scheme; he found that it 'so limited a number of effective classes as to make definitive spatial study of social areas difficult beyond a rudimentary level' (Cowlard, 1979, 240). His response was to subdivide each of the primary classes into three sub-classes, 'a', 'b' and 'c' in order of decreasing status, so that class I became Ia, Ib, Ic, class II IIa, IIb, IIc, and so on. These other schemes have their particular attractions but in the long run the advantage of comparability gained from the already widely adopted Registrar-General's classification, as modified by Armstrong, is a compelling reason for its use.

The socio-economic data obtained from the censuses may be plotted at a variety of spatial scales. The most straightforward approach is to compile information for enumeration districts, that is, for those districts within which individual enumerators collected completed census schedules. The boundaries of the districts are set out reasonably clearly in the enumerators' books. The following extract taken from the 1871 census for the town of Tredegar in the parish of Bedwellty, Gwent, is a typical description of a district:

> That part of Tredegar comprising Morgan Street, Morgan Lane, Circle, Bridge Street, Mill Manager's House, Iron Row, Shop Row, Cross Row, Bedwellty House, Upper and Lower Lodges, Gamekeeper's House, Stables and Price's House by Railway.

Table 2.1 A scheme for the social stratification of household heads from Census enumerators' books (after Royle, 1977, 217)

Class	Qualification
I	Heads whose households
	1 employed more than 25 people
	2 contained at least one servant per household member
	Heads of professional occupation whose households contained at least one servant per three household members
II	Heads whose households
	1 employed between one and 24 people
	2 contained at least one servant per three household members
	Heads of professional occupation
III	Heads whose households contained servants
	Heads of non-manual occupation, including those engaged in commerce
IV	Heads of skilled manual occupation
V	Heads of unskilled manual occupation

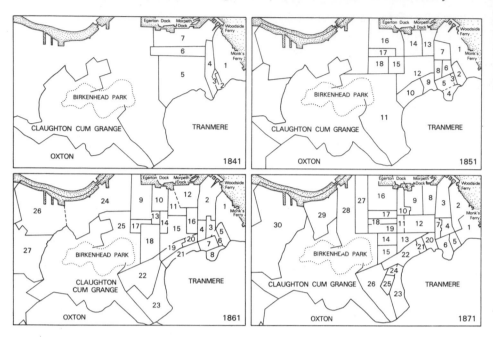

Figure 2.2 Enumeration districts of Birkenhead, 1841–71.
(after Lawton, 1978, Figure 14, 125)

Unfortunately, there are a number of problems associated with enumeration districts. On the one hand, their boundaries were adjusted over time, even from one census to the next, and thus it is difficult to make direct temporal comparisons within towns. By way of example, Figure 2.2 shows the changes in Birkenhead over the period 1841–71. On the other hand, they show considerable variety in size and shape, and may contain rather heterogeneous demographic and social characteristics. Instead of aggregating and mapping the data by enumeration districts, another possibility is to plot variables by grid squares, as was done, for instance, by Shaw (1979) in Wolverhampton and Carter and Wheatley (1982) in Merthyr Tydfil (see Figure 2.3). A grid has the advantage of uniform spatial cells, but the larger the cells the greater the risk of blurring the underlying social patterns. Instead of enumeration districts and grids, both of which are arbitrary spatial divisions, census data may also be presented at the finer scales of streets, rows and courts of houses, and at the level of individual houses if they can be located accurately on base maps. The advantage of working at these finer scales is that they pick out the subtle variations and juxtapositions which characterized tightly packed residential districts.

Rate books

While the decennial censuses are the main sources of social information, the abstraction of data from them involves a great deal of time and effort. For the purpose of identifying broad intra-urban social contrasts, perhaps as a preliminary study to determine sample areas for later in-depth investigations, rate books can be suggested as an alternative. Indeed, the information contained in rate books has been used to analyse urban social patterns, property ownership, migration and, to a

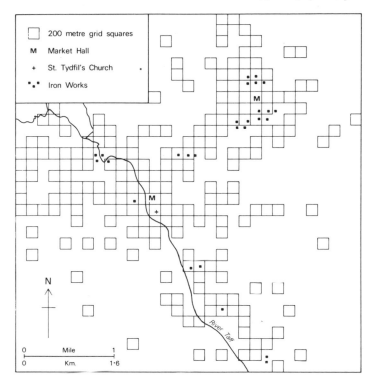

Figure 2.3 Plan of the grid system used in the analysis of census data for Merthyr Tydfil in 1851.
(after Carter and Wheatley, 1982, Figure 4b, 7)

very limited extent, commercial structure. There is the drawback that for some towns the coverage is patchy, but it is not uncommon to find almost complete runs of rate ledgers back to the early nineteenth century.

The basic layout of the ledger pages in the nineteenth century was as shown in Figure 2.4; for each property, in addition to its address and its gross and net rateable values, the sheet includes the names of its owner and occupier and a brief description of its use. For the determination of broad social areas the gross rateable value of a property may be taken as an indicator of the social status of its occupant. This is a crude surrogate measure, but a number of studies have confirmed its suitability. As Robson has observed, 'though the associations between social class, income and rating values are not simple, rating values would appear to be a useful approximate index of class' (Robson, 1966, 121). Using data from the 1961 census, Robson supported his assertion with a correlation coefficient of 0.867 between median gross rating values and social class in the enumeration districts of Sunderland. Subsequent published works, for example, by Gordon on Edinburgh (1979) and Fox on Stirling (1979), have shown the efficacy of the rate valuations in the identification of status areas. Contemporary reports for the nineteenth century, too, argued that the rateable value of a property could be taken as a good indicator of the social status of its occupants. For example, in a report to the Local Government Board on the sanitary condition of the Municipal Borough of Dudley in 1874 it was claimed that 'as a rule the social status and wealth or poverty of the inhabitants of any district may be roughly estimated by the rental of the dwellings they occupy'. The writer, a Dr Ballard, then went on to use rateable values to assess

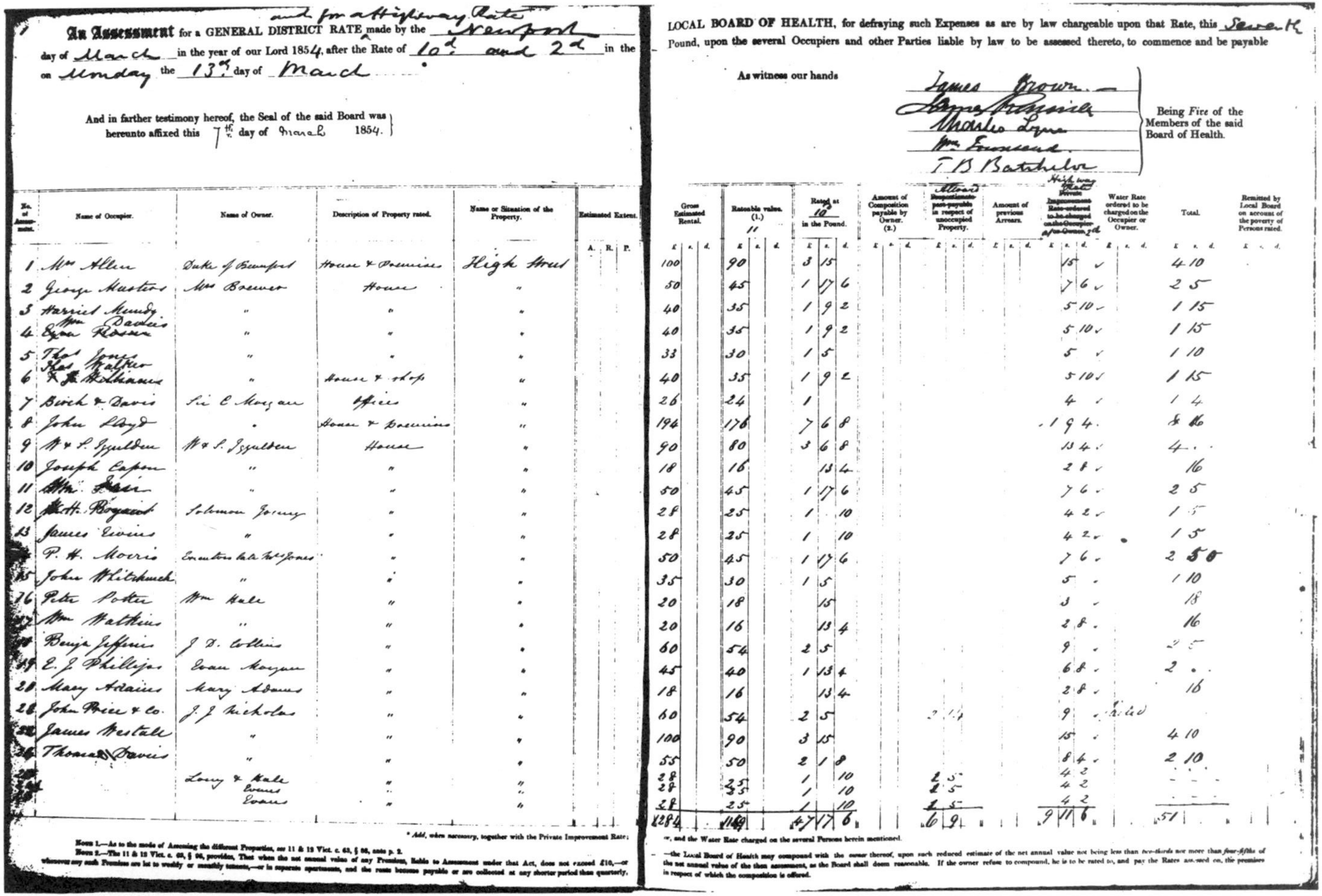

An Assessment for a GENERAL DISTRICT RATE and for a Highway Rate made by the Newport LOCAL BOARD OF HEALTH, for defraying such Expenses as are by law chargeable upon that Rate, this Seventh day of March in the year of our Lord 1854, after the Rate of 10d and 2d in the Pound, upon the several Occupiers and other Parties liable by law to be assessed thereto, to commence and be payable on Monday the 13th day of March.

And in farther testimony hereof, the Seal of the said Board was hereunto affixed this 7th day of March 1854.

As witness our hands

James Brown
[illegible]
Charles Lyne
Wm Evans
T B Batchelor

Being Five of the Members of the said Board of Health.

No. of Assessment	Name of Occupier	Name of Owner	Description of Property rated	Name or Situation of the Property	Gross Estimated Rental £	Rateable value £	Rated at 10d in the Pound £ s. d.	Highway Rate £ s. d.	Total £ s. d.
1	Mrs Allen	Duke of Beaufort	House & Premises	High Street	100	90	3 15	15	4 10
2	George Masters	Mrs Brewer	House	"	50	45	1 17 6	7 6	2 5
3	Harriet Mundy	"	"	"	40	35	1 9 2	5 10	1 15
4	Wm Davies	"	"	"	40	35	1 9 2	5 10	1 15
5	Thos Jones	"	"	"	33	30	1 5	5	1 10
6	Thos Walker	"	House & shop	"	40	35	1 9 2	5 10	1 15
7	Birch & Davis	Sir C Morgan	Offices	"	26	24	1	4	1 4
8	John Lloyd	"	House & Premises	"	194	176	7 6 8	1 9 4	8 16
9	W & S Iggulden	W & S Iggulden	House	"	90	80	3 6 8	13 4	4
10	Joseph Capon	"	"	"	18	16	13 4	2 8	16
11	Wm [illegible]	"	"	"	50	45	1 17 6	7 6	2 5
12	[illegible]	Solomon Jenny	"	"	28	25	1 10	4 2	1 5
13	James Evans	"	"	"	28	25	1 10	4 2	1 5
14	P. H. Morris	Executors late Mr Jones	"	"	50	45	1 17 6	7 6	2 5 0
15	John Whitchurch	"	"	"	35	30	1 5	5	1 10
16	Peter Potter	Wm Hale	"	"	20	18	15	3	18
17	Wm Watkins	"	"	"	20	16	13 4	2 8	16
18	Benjn Jeffries	J. D. Collins	"	"	60	54	2 5	9	2 5
19	E. J. Phillips	Evan Morgan	"	"	45	40	1 13 4	6 8	2
20	Mary Adams	Mary Adams	"	"	18	16	13 4	2 8	16
21	John Price & Co.	J. J. Nicholas	"	"	60	54	2 5	9	[illegible]
22	James Westall	"	"	"	100	90	3 15	15	4 10
23	Thomas Davies	"	"	"	55	50	2 1 8	8 4	2 10
24		Long & Hale	"	"	28	25	1 10	4 2	
25		Evans	"	"	28	25	1 10	4 2	
26		Evans	"	"	28	25	1 10	4 2	
					1284	1149	47 17 6	9 11 6	51

Figure 2.4 Extract from a rate book of 1854 for Newport, Gwent. The left-hand page includes details of each property; the first two columns of the right-hand page show the gross estimated rental and rateable value of each property.
(Gwent County Record Office)

the status of districts in the town (Dr Ballard's report to the Local Government Board on . . . Dudley, 1874). As with census data, analysis can be undertaken at a variety of scales, for arbitrary zones, streets or individual premises. To illustrate their utility at the most detailed level, Figure 2.5 shows the distribution of individual premises with rateable values of £20 and over in Newport, Gwent, in 1880; in addition to highlighting the central business district in High Street, Commercial Street and Commercial Road, the high-status terraces and newly constructed suburban villas are clearly identified (Lewis, 1985a).

While rate books offer a fairly quick means of determining the broad social and commercial lineaments of towns, they compare unfavourably with the enumerators' books in view of the latter's greater detail on households. However, they come into their own on property ownership and occupancy for they specify the

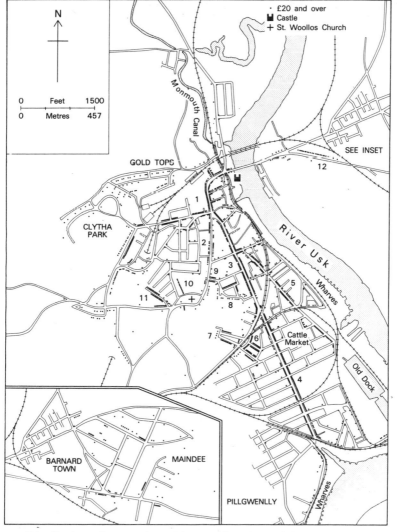

Figure 2.5 Distribution of properties with rateable values of £20 and over in Newport, Gwent, in 1880.
(after Lewis, 1985a, Figure 4, 140)

Table 2.2 Concentration of ownership, Ramsgate, May 1851 (all figures are cumulative) (after Holmes, 1973, 245)

	Properties	Owners	% Properties	% Owners
1 or more properties	1918	650	100.0	100.0
2 or more properties	1570	303	81.9	46.6
3 or more properties	1360	198	70.9	30.5
4 or more properties	1165	133	60.7	20.5
5 or more properties	1037	101	54.1	15.5
6 or more properties	932	80	48.6	12.3
7 or more properties	824	62	43.0	9.5
8 or more properties	747	51	38.9	7.8
9 or more properties	635	37	33.1	5.7
10 or more properties	573	30	29.8	4.6
15 or more properties	428	18	22.3	2.8
20 or more properties	289	15	15.1	2.3
25 or more properties	228	7	11.9	1.1
30 or more properties	173	5	9.0	0.8
35 or more properties	110	3	5.7	0.5
40 or more properties	40	1	2.1	0.1

name of the occupier and the name of the owner of each premises. Although there are some inherent problems in the data, such as the difficulty of distinguishing between individuals with identical names, Holmes has shown from his study of a rate book for Ramsgate for 1851 that it is a reasonably straightforward task to produce tallies of the number of properties owned by individuals (Table 2.2) and to determine the numbers living in premises with owner-occupied and rented tenures (Holmes, 1973). These data can be obtained by drawing up and comparing lists of names from the 'owner' and 'occupier' columns of the ledgers.

The names and addresses of occupiers in successive ledgers may also be used to establish the extent of migration. A deficiency of enumerators' books for migration studies is their inability to pick out the short-term, short-distance moves of many

Table 2.3 Tenancy of Miniature Row, Ramsgate: May 1851 to May 1853. None of the houses were owner-occupied and were all owned by the same landlord (after Holmes, 1973, 249)

May 1851	*November 1851*	*April 1852*
1 Empty	1 Ann Robinson	1 Ann Robinson
2 Henry Rigden	2 Empty	2 James Bussey
3 Empty	3 Empty	3 Empty
4 William Packer	4 Empty	4 Empty
5 Empty	5 Henry H. Evans	5 Henry H. Evans
6 Empty	6 Stephen Norris	6 Stephen Norris

July 1852	*December 1852*	*May 1853*
1 Thomas Jones	1 Thomas Jones	1 George Ratcliffe
2 Empty	2 Thomas Jones	2 Henry Rigden
3 Sarah Edwards	3 Empty	3 Empty
4 Edward Paine	4 Edward Paine	4 Edward Paine
5 Empty	5 Thomas Watson	5 Thomas Watson
6 Stephen Norris	6 Empty	6 Mary Ann Hall

families, unless, of course, each move was between separate communities and coincided with the birth of a child. Again, Holmes has shown that after making due allowance for deaths by consulting local Death Registers, the very large turnover of residents in individual properties can be reconstructed as exemplified in Table 2.3. Since separate rate assessments were often made for every quarter of every year, where good runs of the ledgers have survived it is possible to build up a moving picture of the extent and pattern of intra-urban circulation. Given the column headed 'description of property rated', it might be thought that the same sort of detailed analysis could be carried out for business premises, on their type, persistence and ownership. Unfortunately, the descriptions used by the ledger clerks are vague or unreliable. Properties which other sources confirm as having been shops, are not identified as such in the ledgers; all too often the entry 'house' is all that is given, and at best there may be nothing more than 'house and shop' or 'house and premises'. Apart from providing some useful information on vacancies (Davies, Giggs and Herbert, 1968), rate books are not a particularly fruitful source of precise commercial data.

Directories

Taken together, the enumerators' books and rate books facilitate the analysis of the social, mobility and tenurial characteristics of households, but for commercial and industrial information it is necessary to turn in another direction, to trade directories, which stand out as a prime resource. The nineteenth century saw the proliferation of both national and local directories, and these have survived in large numbers. Shaw's recent guides to directories published in Britain provide useful summaries of their scope, content and coverage and can be recommended as basic introductions (Shaw, 1982; Shaw and Tipper, 1989). Used in conjunction with the census enumerators' books they help to fill out the business particulars of those described in such general terms as 'shopkeepers' and 'dealers', but on their own many contain enough information to permit the analysis of both intra-urban commercial structures (for example, Shaw and Wild, 1979) and inter-urban hierarchical patterns (for example, Lewis, 1975).

At their best, they provide reasonable coverage of towns on a systematic street-by-street basis, showing the name and type of business performed in each building, or, for residential premises, the occupation of the head of household. In addition to these inventories for individual streets, which are usually arranged in alphabetic order, it is quite common for directories to present the same information in two other complementary lists, the one being an alphabetical list of residents with their addresses and occupations, and the other an alphabetical list of trades and professions, with the names and addresses of those engaged in each of them. As can be appreciated from the example in Figure 2.6, these lists are a valuable source of data.

However, it is necessary to add a note of caution over their use. Many writers have pointed to various inaccuracies which crept into their initial compilation and subsequent revision (Lewis, 1975; Shaw, 1982). Comparison between directories produced for the same year by different publishers has revealed significant disparities, as has comparison between directories and census returns. It can be seen from Table 2.4 which compares three directories with each other and with the 1891 census, that the most comprehensive directory contains only 65 per cent of census households. Further, over and above the fact that directories were only concerned with householders, there is clear evidence that there was a bias towards the relatively better off in nineteenth-century society. This bias has been illustrated by

Broadway *continued.*
175 Evans Lewis
177 Hughes Henry
179 Gilbert F., greengrocer
181 Fowler George
here cross over
200 Gillard Samuel, *Royal Oak htl.*
198 Clarke John, coal trimmer
196 Courtney W.
194 Smith John
192 Dando John
190 Cleverdon Thomas
188 Morgan Noah, *New Dock inn*
——*Blanche Street intersects*
182 Noall Henry T., master mariner
180 Peart Arthur W., manager
178 Reed John, engineer
176 Cottrell Robert, mason
174 Cottrell John, dairyman
172 Perry William
170 Gibbs J. C., general shop
——*Arthur Street intersects*
168 Smaldridge John, gen. dealer
166 Watkins Henry, plasterer
164 Cook Charles, cab proprietor
162 Taylor John, mason
160 Edmunds William
158 White Joseph
156 Gibbs William, grocer
——*Maud Street intersects*
154 Williams A., baker
148 Bernard Henry, com. traveller
146 Matthews Frederick, decorator
144 Haddon Isaac, boiler maker
142 Jones Thomas, coal dealer
140 Cawson Ann, grocer
——*Theodora Street intersects*
138 Bowen John & Sons, grocers
136 Sanders William, butcher
134 Ackland S., greengrocer
132 Harris Henry, shoemaker
130 Pincombe William, builder
128 Lucas Benjamin, baker
126 Martin Frederick, stationer &
Broadway Post Office
——*Harold Street intersects*
120 Sanders Thomas, news agent
112 Evans Albert J., canvasser
110 Lowe Thos., *Bertram hotel*
——*Bertram Street intersects*
108 Moore Wm., ironmonger
106 Burns E. J., Prudential ins. agt.
102 Davies & Slee, general dealers
100 Purnell Robert, Customs
98 Lloyd Henry
96 Gaze Albert, Manor farm dairy
94 Jones David, chemist
——*Cecil Street intersects*
92 Jones Samuel, draper
90 Jones Thomas, confectioner
88 Spear Edgar, butcher
86 Evans Sarah, general shop
82 Brown Bartholomew
80 Taylor Joseph, greengrocer
——*Helen Street intersects*
76 Shibko Abraham, pnbkr. & clthr.
74 Shibko Abraham, bt. & sh. wrhse.
68 Dhenin Lewis, baker
66 Taylor William, tinman
62 Alden George, *Locomotive inn*
——*John Street intersects*
58 Harper Charles, coal trimmer
56 Denham T. H.
54 Weeks William
52 Talbot Henry, confectioner
50 Harpur J., grocer
48 Thompson James, baker
46 Carter J., wardrobe dealer

42 Baker F. R., hairdresser
40 Lovell A. J., paperhanger
36 Roberts Richard, boilermaker
34 Eynon William, shipwright
BROADWAY WESLEYAN CHAPEL
30 Surcombe S., coal inspector
28 Love Charles, mason
Love Miss, dressmaker
22 Baker J., dairy
20 Roch J. O., draper
18 Kidney J., bootmaker
16 Summers James, undertaker
14 Mewton S. J., machine dealer
12 Johns Edwin, painter
10 Marks John, toy and fancy shop
8 White Job, butcher
6 Davies Henry, labourer
4 Edwards Richard, wagon maker
Clifton Hotel Wine Vaults

Bromfield Street, *Grangetown*

Holmesdale St. to Bromsgrove St.

1 Slack John, fitter
3 Owen Thomas, pilot
5 Sparks Charles, mariner
7 Pimm Henry
9 Alexander Thomas, coal trimmer
11 Lewis Noah, stoker
13 Bracher Mrs., midwife
15 Leeson George, seaman
17 Williams David, mason
19 Adams Edwin, foreman
21 Scott Mrs. A.
23 Seymour John, engineer
25 Seymour William, engineer
here cross over
26 Caville Henry
24 Martin John, engineer
22 Young Susan
20 Davies John, shipwright
18 Evans John, plumber
16 Cazenave John, tailor
14 Jenkins David, ship carpenter
12 Sage George, seaman
10 Thomas John M., shipwright
8 Meyer John, boilermaker
6 Stiff John
4 Mabbs James, marine engineer
2 Cavill James, engineer

Bromsgrove Street, *Grangetown*

Clive Street to Holmesdale Street

1 Pritchard Henry, grocer
3 Simmons James, engine driver
5 Symes Emma, tailoress
7 Donaldson John, rigger
9 M'Connell John, foreman painter
11 Goode John, mariner
13 Gower Daniel, master mariner
15 Careless Robert, general dealer
17 Gristock James
19 Mardon William
21 Clare Harriett
23 Hutchins M. G. & E., drapers
ST. PAUL'S CHURCH.
25 Grant Thomas, railway foreman
27 Wilke Thomas, engine driver
29 Fisher Hy., contractor's foreman
31 Miller Frederick, dock gateman
33 Wilkes William, engine driver
35 Jones Alfred, mariner
37 Harris Charles, hobbler
39 Bristow Charles, ship's steward
41 Evans William, crane driver
43 Redman Henry
45 Richards James, storekeeper

47 Radford Francis, carpenter
49 Crouch Charles, mariner
51 Smith Robert, foreman
GRANGETOWN BOARD SCHOOLS.
Kent Edward, master
Butterworth Mrs., mistress
Gowing Miss, infants
here cross over
14 Howe John A., pilot
12 Wiltshire Robert, fish & fruit dlr
10 Ford Lewis, shipwright
8 Walkom John, coal dealer
6 Abraham Charles, pawnbroker
4 Parry David, blacksmith
2 Saville John, mariner

Brook Street, *Canton*

Green St. to Coldstream Terrace

1 Watson Mackenzie
3 Evans E., *Coldstream hotel*
5 Lewis John, coal trimmer
7 Ashton Roger, fishmonger
9 Maskell Mrs.
11 Service Isaac
13 Cockin Henry
15 Hall J. M.
17 Cole Aaron, plumber
19 Creece Leontine F., painter, &c.
21 Phillips Miss Ann
23 Thomas Thomas, builder
25 Ford John
27 Rawlinson George, hairdresser
29 Webber Edwin
31 Martin William, fitter
33 Thomas Edward
35 Sweetman N. W.
37 Evans James Richard, carpenter
39 Fowler Joseph, carpenter
41 Rees Mrs. Mary Ann
43 Rees Daniel, coal trimmer
45 Williams Frederick
47 Laird George, engineer
49 Organ George
51 Greenman William
53 Morgan David, haulier
here cross over
44 Rogers John, smith
42 Brown John, traveller
40 Lucock Edwin, manager
38 Rusden Philip
——*Plantagenet Street intersects*
36 Harvey William
34 Harris William, plumber
32 Nichols James, carriage inspector
30 Barrett William, engineer
28 Beer John, milkman
26 Williams Mrs.
24 Hill Charles A., assistant
22 Lewis John
20 Scott William, engineer
18 Cortens Peter, tailor
16 Calaghan Mrs.
14 Bicknell William, tailor
12 Lock John, butter dealer
10 Ware Henry
8 Sievwright Mrs.
6 Evans Lewis, mariner
4 Davies Benj., wagon inspector
2 Lewis Mrs.

Brunel Street, *Canton*

Eldon Street to Craddock Street

BAPTIST CHAPEL
1 Williams William, Moat house
3 Julian David, pilot, Hendre hse.
here cross over

Figure 2.6 Extract from D. Owen and Co's (Wrights) Cardiff Directory for 1890. (National Library of Wales, Aberystwyth)

Table 2.4 Variations in directory coverage in Exeter, by type of publication (after Shaw, 1982, 33)

Publisher/Date	Number of names	% of households listed in 1891 census
White, 1890	5401	65
Kelly, 1889	4817	58
Besley, 1890	4244	51

Shaw, drawing on information for Ashby-de-la-Zouch compiled by Page from Whites directory of 1862 and the census enumeration books for 1861: 'For example, in the census 35 per cent of Ashby's households were employed in craft activities, and 18 per cent as tradesmen. In contrast, 33 per cent of the names listed in Whites directory are tradesmen, and only 25 per cent are classified as craftsmen. Similarly, labourers formed 7 per cent of the households enumerated in the census for Ashby, whilst the directory did not list any labourers or domestic servants' (Shaw, 1982, 34).

Despite these and other deficiencies, the trade lists still stand up as worthwhile sources, particularly for studies of commercial and industrial activities. It seems that directories are at their most reliable for the main thoroughfares which housed the majority of retail and professional facilities. They are more than just a supplementary source to the census enumerators' books: they have a longer run than the census, they are not subject to a confidentiality rule, and many contain extra descriptive material on local history and topography, transport, administration and commerce.

Perhaps it is appropriate to mention here a useful supplementary source of information on the commercial structures of many of Britain's larger towns, the Goad Fire Insurance Plans. These detailed maps, produced by the Charles E. Goad company for insurance assessment purposes, recorded the constructional and land-use characteristcs of premises in and around central business districts. They were not confined to shops and offices but often included warehouses and factories, particularly those with a high fire risk. Based on field survey and updated at fairly regular intervals (generally every five years or so, but more frequently for larger places), they were first produced within the British Isles in 1886, and by the end of the century the central parts of the major towns and cities had been completed. Although at their best for the twentieth century, for really detailed work on the spatial and vertical arrangements of business premises within centres, the extant plans of the late nineteenth century offer a potentially fruitful source of data. Certainly they contain much more precise locational information than the average directory.

To date, this information has not been widely used, largely because the plans had not been systematically examined and catalogued. Recently, however, this task was put in hand by Rowley, and he has published a good introductory statement on the content and coverage of the Goad maps and documents (Rowley, 1984). Undoubtedly, their late appearance and limited coverage in the nineteenth century are disadvantages, but such is their detail that they may help to answer questions about the changing spatial distributions of different types of retail, office, warehousing and manufacturing businesses, and the amount of residential occupancy of the upper floors in commercial premises, in specific towns. Rowley's inventory shows that between 1886 and 1900 over 40 separate urban areas were surveyed (and in some cases revised as well), including ports, resorts, regional markets and manufacturing towns.

Building plans and registers

Turning from the occupants of urban premises to the built form itself, one means of determining the structural features of new property, and of identifying those involved in its construction, is by examination of extant building plans and registers. These proliferated in the second half of the century as more and more urban authorities insisted that new premises should conform to building by-laws. The procedure was quite simple. Builders were required to submit their proposals to local councils for scrutiny. There was no standard format for these building applications but most authorities required at least the basic essentials of a plan and a brief description of the proposed development.

> When a plan was submitted to an authority a brief description was usually placed in a register, naming the number and type of buildings, the street in which they were to be built, and the building-owner, the latter sometimes being accompanied by, or replaced with, the name of the builder or simply of an agent. For example, an entry from the Leed's *Index of Plans, Building Inspector's Office* for 1896 records:
>
>> Ash Road, Grimthorpe Street and Back Grimthorpe Street (*Name of street*), Headingley (*Township*), William Gibson (*Name of owner*), 4 through houses (*Description of property*), 36 (*Folio no.*), 20th March 1896 (*Date of approval*), ------- (*Date of completion*)
>
> At certain times, fortnightly or once a month, a committee met to consider these plans and its decision was recorded (Aspinall, 1978, 3).

In some cases the dates on which construction was commenced and completed were also entered.

Preliminary surveys of local archives have shown that these documents have survived in large numbers for many towns (Aspinall, 1978; Aspinall and White-hand, 1980). Although it is difficult to determine the precise roles of those described as builders, owners and developers, the registers do enable analysis of the sizes of building firms, the duration of projects and the chronology of urban expansion, plot by plot and street by street. Further, the plans themselves contain valuable information on the construction, layout and facilities of houses and other premises which have long been demolished.

Descriptive and statistical accounts

The sources discussed above facilitate the reconstruction of broad intra-urban patterns. However, to derive more detailed information on the characteristics of and conditions in particular sub-areas it can be rewarding to turn to a variety of descriptive accounts of towns or parts of towns, ranging from individual and perhaps highly subjective impressions to the official reports of commissions and committees. As Dennis has shown in his recent summary of contemporary accounts of nineteenth-century cities, the range of material is very wide indeed (Dennis, 1984).

Although their descriptions were sometimes presented in rather flowery language, many individuals made quite perceptive comments on the changing social fashions and the changing spatial patterns in towns. Some were able to draw on their experiences in more than one of the country's major cities, whereas others were much more parochial in their observations. Engels, for example, in his book on the condition of the working class in England in the 1840s, displayed knowledge of London, Manchester and other great cities, and highlighted the main geographical patterns within them, as shown by the following extract taken from his description of Manchester:

> Manchester contains, at its heart, a rather extended commercial district, perhaps half a mile long and about as broad, and consisting almost wholly of offices and warehouses. Nearly the whole district is abandoned by dwellers, and is lonely and deserted at night; only watchmen and policemen traverse its narrow lanes with their dark lanterns. This district is cut through by certain main thoroughfares upon which the vast traffic concentrates, and in which the ground level is lined with brilliant shops. In these streets the upper floors are occupied, here and there, and there is a good deal of life upon them until late at night. With the exception of this commercial district, all Manchester proper, all Salford and Hulme, a great part of Pendleton and Chorlton, two-thirds of Ardwick, and single stretches of Cheetham Hill and Broughton are all unmixed working-people's quarters, stretching like a girdle, averaging a mile and a half in breadth, around the commercial district. Outside, beyond this girdle, lives the upper and middle bourgeoisie, the middle bourgeoisie in regularly laid out streets in the vicinity of the working quarters, especially in Chorlton and the lower lying portions of Cheetham Hill; the upper bourgeoisie in remoter villas with gardens in Chorlton and Ardwick, or on the breezy heights of Cheetham Hill, Broughton, and Pendleton, in free, wholesome country air, in fine, comfortable homes, passed once every half or quarter hour by omnibuses going into the city. (Engels, 1969, 79).

However, many local commentators were equally aware of trends and patterns in their own communities and wrote about them in town histories and guides. Similarly, local newspapers were replete with descriptions of residential districts and their inhabitants, and of changes in industry and commerce. Further insight into the actual experience of living and working in a particular town or locality can come from literature, both autobiography and fiction. Dickens's description of Coketown and Bennett's account of Bursley are classic examples (Dennis, 1984, 92–6). In general, though, while giving the feel for people and places, this material is not so amenable to detailed geographic analysis.

In addition to these personal accounts of towns there is a mass of descriptive and statistical material contained in 'official' documents. Particularly active were the various bodies and agencies concerned with public health and housing which were both the instigators and recipients of numerous national and local reports into urban conditions. At the national level, tucked away in the papers of committees and commissions which reported to Parliament are descriptions of various aspects of town life, some of which contain valuable material on intra-urban residential and working conditions. One way into this mass of information is through the *Catalogue of British Parliamentary Papers* (1977) which was published by the Irish University Press to accompany its facsimile reproduction of reports and evidence. The Irish University Press published the documents in subject sets – nine volumes on Municipal Corporations, thirty volumes on Poor Law, three volumes on housing, ten volumes on planning, seventeen volumes on health, and so on – and many of these sets contain useful descriptive and statistical information. Not only do they provide summaries of conditions in towns in general, but also supporting evidence for individual places.

Among the first of the major national surveys was the Chadwick report for the early 1840s on the sanitary condition of the labouring population in England, Wales and Scotland. This was followed by successive commissions and committees which received evidence on health matters in individual towns, for example, the Royal Commission on the state of large towns and populous districts, 1844–45, and the Royal Commission on sanitary laws, 1868–71. Over the century, similar committees dealt with a variety of other urban deficiencies and issues which caused public concern, including overcrowding, slum clearance, the provision of working-class housing, tenure and employment.

Running parallel to the publication of these national surveys was the preparation of local health and housing reports. Of particular note are the descriptions of individual towns which were submitted to the General Board of Health and its successor the Local Government Board in the second half of the century, following the Public Health Act of 1848. Without doubt many of these contain valuable information. As an illustration of the utility of their content for geographical studies, one example has been selected, part of a report of 1874 on Wolverhampton. In it, Dr Ballard identified the main social divisions within the town:

> The differences between the western and eastern sub-districts . . . are connected with a difference in the character of the populations inhabiting them, and especially in the character of the population which, since 1861, has constituted the increase in each. In the western sub-district, which is that most remote from the coal-field, the most well-to-do part of the population resides, and the families include a considerable proportion of female domestic servants . . . The eastern sub-district, on the contrary, is chiefly inhabited by that part of the population which is engaged as work-people in the mines, forges, and manufacturing establishments, which are chiefly located in or near this part of the borough. The distinctions thus brought into notice are important, in as much as they will assist in explaining differences observable in the rate of mortality in the two sub-districts, and especially in the rate of infant mortality . . . The eastern half of the town is constantly smoky, being that part in which most of the manufacturing establishments are situated. The atmosphere of the western half, however, is generally pretty clear, since the prevailing winds, south-west and north-east, pass over an agricultural country (Dr Ballard's report . . . of Wolverhampton, 1874, 2).

In addition to these surveys for the national Boards, the annual reports of Medical Officers of Health which were prepared for their own Local Boards of Health also warrant attention. Not only do they deal with health matters but they often highlight socio-spatial contrasts. A good example is that prepared by H. J. Paine for Cardiff in 1855. He went as far as to subdivide the town into five distinctive geographical regions, the North, East, West and South and Newtown, and to describe conditions in each. The North, for instance,

> comprises that portion of the town which extends from Crockherbtown to Castle street; it includes all streets opening into this line, being bounded on the south by Quay street, Church street, continued on through Ebenezer street to Charles street. It is upon a higher level than the other portion of the town. It is inhabited principally by gentry, professional men, and respectable tradesmen. It contains but few courts, these being in good condition, both pitched and paved.

In contrast, the Newtown district

> has a low level; the streets have recently been pitched; the houses for the most part are occupied as Irish lodging-houses, and are seriously overcrowded. (H. J. Paine, Third annual report, 1856, 4–5)

Given the number and variety of local reports, they can prove to be very useful sources of descriptive and statistical information on individual towns.

Maps

Last, but by no means least in terms of their worth in geographical studies, are maps and plans. At the most obvious level, base maps are required to plot the data derived from the standard sources discussed above. However, over and above this use, maps and plans contain information in their own right, on the chronology, layout and density of development, and sometimes on land use and tenure. Quite a lot has been written on the availability and content of historical maps in Britain, and thus only a bare outline need be given here. The details can be obtained from the existing guides to map sources, such as those prepared by Harley (1972) and Harley and Phillips (1964).

Prior to the entry of the Ordnance Survey into the field of urban mapping in the middle of the nineteenth century, town plans were produced by private enterprise. Some cartographers, such as William Fairbank who mapped Sheffield in 1771 and William Green who mapped Manchester and Salford in 1794, were local land surveyors, but others, such as John Rocque in the middle of the eighteenth century and John Wood in the early nineteenth century, were engaged in much more substantial map-making businesses and gained national reputations. A cursory examination of the entries for individual towns in *The British Museum catalogue of printed maps, charts and plans* (1967) reveals that large numbers of private surveyors, cartographers and engravers were at work across the length and breadth of the country. As can be appreciated from the extract from a John Wood map reproduced in Figure 2.7, many of their plans are highly valued for their detail and clarity. Inevitably, in the absence of a national mapping agency, the coverage of individual towns before the 1850s is patchy. While some places have a good run of fairly detailed plans from the early 1700s, others are much less fortunate.

Perhaps the first maps which can be described as national in scope, and which are at a scale large enough to show separate premises, are the Tithe maps. These parish or township surveys were produced for the Tithe Commissioners between about 1836 and 1851. They comprise two documents, a plan, on which parcels of land and buildings are numbered, and an accompanying written key or schedule, showing

> standard information as to the owner, the occupier, the name and description of the
> lands and premises, the state of cultivation, the acreage of each parcel, and the rental
> apportioned on the land within the titheable area. The total result is an immensely
> detailed – almost exhaustive – local inventory. (Harley, 1972, 35)

Clearly, these surveys are most useful for studies of land use and tenure in rural districts but they can be employed in urban research: some of the plans and schedules include built-up urban land, but they are of greatest use in looking at the release of agricultural land to the building industry as towns pushed outwards into their rural surrounds.

Almost coincidentally with the completion of the Tithe surveys, the Ordnance Survey commenced the publication of large-scale maps and plans. These are the main sources for the second half of the century. At the national level it produced the six inches to one mile (1:10,560) and the 25 inches to one mile (1:2500) series, the latter showing the layout of individual properties within urban areas. Showing more detail still, it published town plans at the scales of 1:1056 (five feet to one mile), 1:528 (ten feet to one mile) and 1:500 (10.56 feet to one mile). By the 1890s towns with over 4,000 inhabitants had been surveyed at one or more of these three scales. The great advantage of these detailed plans over the twenty-five inch for geographical studies is the inclusion of more written information, particularly the names of buildings and the uses of some commercial and industrial premises. Not only is this information of value in its own right to locate particular premises with particular functions, but the named buildings provide vital reference points for plotting the data given in other sources. Figure 2.8, which is part of a 1:500 plan of Welshpool published in 1885, illustrates the detail and clarity of the large-scale sheets.

Figure 2.7 John Wood's plan of Aberystwyth, 1834. In addition to showing the basic layout of streets and properties, this detailed plan contains some information on land ownership and identifies most of the prominent buildings in a numbered key.

Decision-makers

All of the sources discussed above facilitate the mapping and description of the changing spatial patterns in Britain's towns over the course of the century. They may be supplemented by a variety of other potentially useful documents which are not examined here, but which are discussed in Stephens's inventory of resources for local historians (1981). However, in order to investigate the processes which lay behind those patterns it is necessary to turn to other sources. By and large, the most fruitful of these are the personal and business records left by those who lived and worked in Britain's towns, records which contain information on the decisions made by landowners, builders, householders, businessmen and so on in their

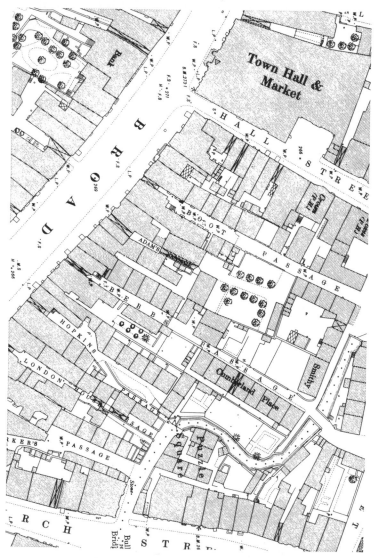

Figure 2.8 Extract from the 1:500 plan of Welshpool, Powys, published by the Ordnance Survey in 1885.
(Department of Geography, University College of Wales, Aberystwyth)

day-to-day activities. It was the convergence of innumerable individual, yet similar, decisions which created the broader, or aggregate, social, commercial and morphological divisions on the ground.

This need for information on the decisions made by individuals is not easily satisfied for, inevitably, the survival of personal records has been patchy, spatially, temporally and across social classes. For any one town, information must be built up by painstaking research of family and business papers, and even then only fragments of empirical evidence may be obtained on some of the decision-makers. In the case of the release of land for urban building, for example, at best a large landowner may have left the legacy of copious estate papers detailing the negotiations which were conducted with agents, developers and builders, and thus there is every possibility of getting at the decisions and the reasons for them; at worst, though, land may have been in the hands of numerous small owners, none of whom have left any record of their property transactions.

Then again, in the case of householders, there is fragmentary material on the reasons for living in particular localities, and for moving house. While the papers of the well-to-do, the best educated and the most literate members of society give some insight into the motives and aspirations of the middle classes, there is a paucity of information on the lower ranks. Some use has been made of personal diaries and journals to disentangle the motives and spatial moves of individuals and families, but these are more in the nature of 'lucky finds' in record offices than standard data sources. However, where they do exist, they are valuable. One such example is the diary of David Brindley which has been used by Lawton and Pooley to examine residential mobility in Liverpool over the period 1882–90; even here, though, it should be noted that the record is stronger on the spatial pattern of moves than it is on the motives and decisions (Lawton and Pooley, 1974), and there is no means of assessing how representative Brindley's moves were. It might well be that only the exceptional kept and preserved such a diary.

Over and above the decisions of individuals, some aspects of town life were shaped by corporate bodies, such as the municipal and town councils themselves and the boards concerned with the supply of public utilities. The extant papers of these bodies, especially the minutes taken at meetings, help in the task of unravelling the decisions which were made. As Stephens has commented, the most useful records of meetings 'may give details of the debates and touch on the reasons behind the decisions taken'. Such was the breadth of topics discussed in town councils that their minutes can be particularly helpful:

> These records are an important source of information not only on the working of local government but on every aspect of town life, political, social, religious, and economic. Evidence is to be found on the relationship between membership of the corporation and the different social, economic or religious groups in the town, the attitude of the corporation to business, trade and industry, its relationship with other towns, with local aristocracy, and with Parliament and the central government. Details of public health, town planning, public services, charities, poor relief, education, law and order, markets and fairs and many other subjects may be discovered in the minutes. (Stephens, 1981, 86)

Given the number and variety of decision-makers it is quite impossible to produce a comprehensive inventory of their records here. All that can be done is to stress the importance of their papers, which range from the private and perhaps anecdotal jottings of individuals to the official records of local councils. Through careful scrutiny of these documents it is possible to go some way towards pinpointing the decisions which were made.

Conclusion

The balance of this chapter has been heavily weighted towards those data which facilitate the reconstruction of spatial patterns. It has set out the strengths and weaknesses of the standard sources which have been used to good effect in geographical research; within that discussion special attention was given to the enumerators' books of the decennial censuses. However, it must be recognized that by their very nature, many of these sources do restrict analysis to the mapping of spatial patterns at successive points in time. In order to get at the processes which created those broader patterns it is necessary to try to uncover what decisions were made, and by whom. Thus, the last part of the chapter has emphasized the need to examine surviving private and public papers which, although fragmentary, contain some answers to questions about the social, economic and political pressures and trends in Britain's towns.

3

The growth of the urban population

Any study of towns in the nineteenth century, whatever its purpose, must surely be based upon a general consideration of the growth of the urban population. But in a geographical context, such a consideration involves two related but somewhat different aspects. The first is concerned with the national increase in numbers, an increase which became so predominantly urban that by 1861 total population increase was occurring at the same time as rural depopulation or loss. The second aspect is the way in which the increased population was distributed amongst the urban centres, both old and newly created, that is with the transformation, if such it was, of the urban hierarchy. Even such a division into two is inadequate for there is a third overlapping aspect, the way in which urban growth during the century was regionally varied so that generally, in terms of urban population growth, and specifically in relation to the evolving urban hierarchy, there were marked regional patterns of growth and decline during the century. Although it is not an easy division to sustain, this chapter will consider the first aspect and the succeeding chapter the second, whilst the third will form part of both.

All that follows in this book deals with the processes involved in, and the consequences of, massive increases in the numbers of people agglomerated in settlements of unprecedented size. Such a statement echoes the threefold basis of Wirth's *Urbanism as a way of life* (Wirth, 1938), the size, density and heterogeneity of populations, and although they have long been regarded as superficial, or superstructural, characteristics as against more fundamental processes, nevertheless in the most direct of senses they constituted the conditions which precipitated the changes with which the latter part of the book deals. But as is implied by calling them 'superstructural' something more is needed than the narration of facts concerning the size, density and heterogeneity of urban populations. The problem is, in a book concerned specifically with urban geography, to choose a path between pure description and the very broadest of interpretations of the developing 'world system'. This choice can be briefly examined.

A first approach can best be defined as descriptive, for it would concentrate on providing an account of the changes which took place and with the generalities which can be derived from observing regularities within the statistics themselves. Into this category would fall the standard pattern analyses of maps which is at the centre of the geographical tradition. A now classic paper is that by C. M. Law (1967). It would take in such attempts at generalization as Ravenstein's 'Laws of Migration' (Ravenstein, 1885).

A second approach would be more analytical, seeking broader principles in order to explain the characteristics of urban population growth. Amongst these would be included that group of associated models which interpret urban increase via the range of demographic factors, as for example in the context of the demographic transition. A parallel in economic terms would be the element of urban change in Rostow's stages of economic growth (Rostow, 1960). Undoubtedly in geographic terms the major British contribution has been that of Richard Lawton (Lawton, 1968, 1983, 1986).

A third and broadest approach is not easy to characterize but it can be called 'structural' although that word does carry a wide range of meanings. Urban growth would now be seen as a local response to a much wider process of fundamental change, and one where the isolation of England and Wales is inappropriate. The critical transformation was that brought about by industrial capitalism which was the organizing principle within an evolving world system, which itself would have to be traced far back beyond the late eighteenth and nineteenth centuries. But the central principle of the capitalist mode of production was the divorce of the labourer from the complete product marketed; the labourer sells not an object or artifact fabricated, or a service, but simply labour to an entrepreneur, who is thus enabled to appropriate a money value through the market system. The machine replaced the workman with a set of tools. In order that the system could operate, where the multiple tasks involved in the making of a complex product were broken down into a series of separate jobs, labour needed to be assembled at one point. That was in the factory, the distinctive creation of the industrial revolution. This was the progenitor of urbanization, the basic imperative to agglomerate labour which, both in extent and location, was a response to capital seeking to maximize returns. Perhaps even more fundamental was the need for control over workers to make sure that work was carried out at the maximum pressure and that the product was uniform. It can be argued that the real drive to the factory system came from such organizational factors rather than from technological demands (Davis and Scase, 1985, 14–15). Davis and Scase contend that 'factories were disciplinary systems since control over work performance was exercised through supervision, by the division of labour and by mechanization' (p. 16). The rest is adventitious. For example, public health legislation during the nineteenth century and the attempt at lowering urban mortality rates would be interpreted as the response to the need to maintain an effective and efficient labour force, that is, to the conditions of the market.

Two difficulties arise from following this last interpretation. The first is that in doing so one becomes involved in the complex nexus of Braudel's (1984) and of Wallerstein's exegesis of a world system (Wallerstein, 1974) which becomes farther and farther removed from the immediate urban geography of nineteenth-century England and Wales. The other consequence is that the processes of change are seen as a part of a very long evolutionary sequence, and the notion of an industrial *revolution* itself becomes questionable. Indeed, it would not be amiss to contend that much recent economic history has been directed to that very end. 'But the "industrial revolution" will not go away . . . some recent calls for its banishment from our class notes and articles notwithstanding . . .' (Mokyr, 1987, 293). Working backwards, faced with the interpretation of quite vast urban change, some form of revolution in the mode of production seems predicated. But here lie the paths of diversion. To move towards an examination of why industrialism began in Britain would be to extend the frontiers of a book on urban geography well beyond sensible confines. It follows that the approach in this chapter is more modest and is confined primarily to the presentation of the facts of urban population growth and their analysis.

The growth of population

At the first census of 1801 the total population was 8,892,526. By 1891, the last census of the nineteenth century, it had risen to 29,002,525, an increase of some 326 per cent. By 1901, the total had reached 35,527,843, giving a 366 per cent increase over the 100 year period.

The major study of the population of England must be that by Wrigley and Schofield (1989) and the models they present suggest that these massive increases were due to the collapse of what they term the classic pre-industrial system. Two of the steps are shown in Figures 3.1 and 3.2, England in the early nineteenth century and England towards the end of the nineteenth century respectively. By the early nineteenth century the positive link between population size and food production which had been critical in previous times had been eliminated and hence no connection is indicated in the diagram, whereas at earlier periods a strong positive link had been indicated. Wrigley and Schofield call that connection 'one of the most fundamental and strongest of all the features of the classic pre-industrial system' (Wrigley and Schofield, 1989, 473). Its breaking eliminated the preventive check cycle which constrained population growth, 'Food prices, though they still had a controlling influence on real wages, which had fallen to a very low level in the early nineteenth century, were no longer affected by high rates of population growth . . . The strikingly close positive link between them, which had been maintained at all phases of the cycle during the preceding 250 years, had vanished' (Wrigley and Schofield, 1989, 475). Henceforth, agriculture was able to lift production until later in the century when imports could supplement home production.

Three other characteristics of the early nineteenth-century model are noted by the authors. The first is the positive feedback loop which connected up real wages, the demand for secondary and tertiary products derived from it and the demand for labour, a loop which was strengthened during the period. The second is the well established link between urbanization and mortality as public health problems were generated in the towns. The third is the uncertain relationship between the

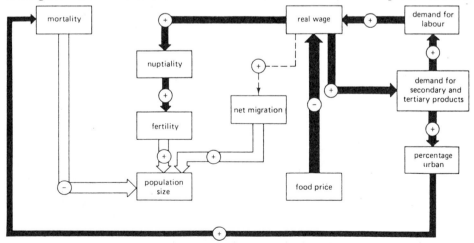

Figure 3.1 England in the early nineteenth century. Connections between demographic and economic factors.
(after Wrigley and Schofield, 1989, 474, Figure 11.8)
Broken lines indicate weak connections, thin lines indicate firm but not powerful connections and thick lines indicate strong and influential links over the whole system.

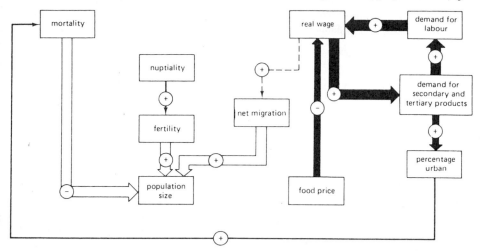

Figure 3.2 England towards the end of the nineteenth century. Connections between demographic and economic factors.
(after Wrigley and Schofield, 475, Figure 11.9)
 The nature of the connecting lines is as on Figure 3.1.

real wage and net migration for presumably economic opportunity was a prime stimulant of immigration, though it is difficult to establish a direct relationship in statistical terms.

By the end of the century 'the system is . . . dominated by the connection between rising real incomes and growth in the secondary and tertiary sectors' (p. 476). Agriculture had become an industry in its own right and hence the logical foundation of the traditional system of demographic regulation had been undermined. The positive link between real wages and nuptiality had disappeared. Inevitably the logical links must remain for population can only be directly related to births, deaths and migration, but all the surrounding empirical links had either disappeared or were greatly reduced. Hence in Figure 3.2 the positive feedback loop of wages, product demand and labour demand predominated with its clear association with the urban proportion.

This interpretation of population change is not far removed from that of the demographic transition, although with a much greater specification of the processes at work. However, a simplistic view of the progressive decline in death rates throughout the century, allied to a birth rate which remained high, cannot be sustained. But in the broadest of contexts it is apposite. True, population increases of considerable magnitude characterized pre-industrial societies but surges of high mortality, the malthusian checks, did operate. It was the diminution and elimination of these surges which generated high and consistent growth. The steady improvement in environmental conditions allied to the transformation of agriculture and food importation maintained low mortality rates, especially among the young. Later parts of this book will highlight the degrading conditions of the poorest areas of towns and stress their high mortality rates, but they were localized and limited in their impact in a century of progressive change.

The initial impact of industrialization seems to have increased fertility rates so that 'a combination of continued high fertility rates with declining child mortality produced the Victorian family, large by legend . . . and probably larger in fact than European families for many centuries' (Wrigley, 1969, 183–184). Table 3.1 shows the fall in fertility in the second half of the nineteenth century as family limitation

Table 3.1 Fertility in England and Wales 1850–1900 (after Wingley, 1969, Table 5.15, 195)

	Live birth rate per 1,000 total population	Live birth rate per 1,000 women aged 15–44
1851–60	34.1	144.9
1861–70	35.2	151.0
1871–80	35.4	153.6
1881–90	32.4	138.7
1891–1900	29.9	122.7

came into operation but it is only in the last quarter that it becomes clearly identifiable. It is against this background that the increase of total population must be set.

Population numbers are a consequence of births, deaths and migration. It is a truism that the larger the scale of analysis the less important is migration. In a closed system migration cannot increase numbers, it can only redistribute them. England and Wales certainly did not constitute a closed system and the major stream of immigration was from Ireland. Between 1841 and 1851 alone the number of Irish born in England and Wales increased from 291,000 to 520,000. Another source was created by the unsettled conditions in continental Europe which engendered a considerable Jewish movement into Britain. But there was a counter-balancing emigration especially to the USA, Canada and Australia.

The growth of the urban population

By this point some general explanation of the massive increase in population in England and Wales has been offered. But the critical issue, both for this chapter and for the book, is the fact that the increase was chanelled into the urban areas to the extent that actual rural depopulation was widespread in the second half of the century. A central problem in the interpretation of urban growth is the inadequacy of census data, especially before 1851, a topic dealt with in the previous chapter. Fortunately the basic data have been provided by C. M. Law (1967). Aware that the administrative basis of the British census was liable to lead to distortions, he recalculated urban populations using three criteria, a minimum population of 2,500, a minimum density of 1 per acre (two of Wirth's conditions) and a measure of nucleation which involved a rather more arbitrary use of personal judgement. Table 3.2 is derived from his figures.

During the one hundred years from 1801 to 1901 the total population increased 3.66 times, but the urban population increased 9.46 times. That increase was greatest in the first half of the century when the urban population multiplied 3.22 times as against 2.2 for the whole of England and Wales; these figures compare with 2.94 times and 2.01 times for the second half. The maximum decadal change was recorded between 1811 and 1821, strangely anomalously at a time when economic historians are greatly concerned at propounding the slowness of economic growth. There was a lower secondary peak between 1871 and 1881. Most significantly by 1851 urban increases were nearly equal to total increases and by 1861 it was considerably greater as actual depopulation became widely characteristic of rural areas. That loss, of course, was due not to natural decrease but to outmigration and that raises the complex problem of the relation of the vital processes, net migration and the nature and regional variation of urban growth.

Four sets of evidence can be used to examine this problem. The first is a relatively simple study by Friedlander which seeks to identify the dates at which counties became urbanized (Friedlander, 1970). There are two critical deficiencies in the study. The first is that the county is a crude area basis, whilst the second is that the date at which the county became urbanized begs all the questions to be asked. Friedlander takes the percentage of male employment in agriculture as his criterion and sets up a figure of 10 per cent in 1951 by a process of intuitive assessment. He then re-works that figure for earlier dates by reference to the structural changes in the distribution of occupations. That process gives figures of 15 per cent for 1851 to 1881 and 12.5 per cent for 1881 to 1921. The series of maps for each decade 1851 to 1951 which Friedlander produced have been incorporated into one map by

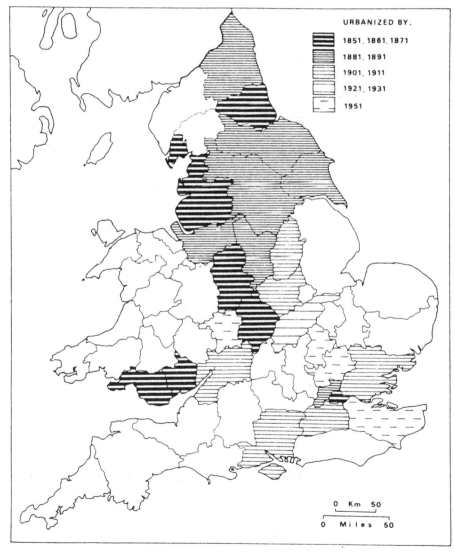

URBANIZED BY:

- 1851, 1861, 1871
- 1881, 1891
- 1901, 1911
- 1921, 1931
- 1951

Figure 3.3 Spread of Urbanization by counties of England and Wales 1851–1951. (after Lawton, 1983, 188, Figure 2 derived from Friedlander, 1970.)
For explanation of the date of urbanization.

Lawton (1983, 188) and this is reproduced as Figure 3.3. The two next pieces of evidence are both from the work of Richard Lawton. The first, included here as Figure 3.4, is his map of population changes by registration districts for the period 1851–1911 in which natural increase or decrease is set against gains or losses by migration (Lawton, 1968, 67). The second is his re-working of a tabulation for natural and migrational change for sets of selected towns, also in the period 1851–1911 (Lawton, 1983, 200). Finally, C. T. Smith's map of migration currents from the 1861 census is reproduced as Figure 3.5 (Smith, 1951, 206). The one feature which stands out in relation to all these four maps and one tabulation is that

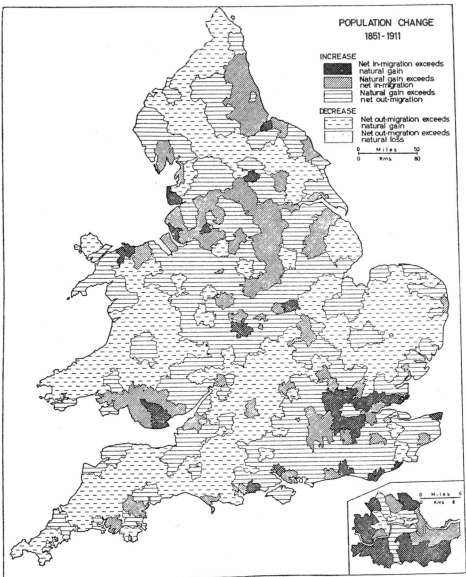

Figure 3.4 Population change in England and Wales 1851–1911.
(after Lawton, 1968, 67, Figure 9)
    Note that not all eight categories possible occur.

they all refer to the period after 1851. This again is a reflection of the availability and reliability of census data on urban populations.

All the evidence indicates that the process of growth was anchored in a series of regional centres. Of these, London was the earliest and the most significant. The universal problem of the relation of actual physical extent to the area for which population was returned makes it difficult to present a realistic assessment but, to follow Weber (1899, 46), the figure of 864,845 was given to the metropolis in 1801 and 4,232,118 was the total for the county of London, created in 1858–59, in

Figure 3.5 Migration currents in England and Wales 1851–1861.
(after C. T. Smith, 1951, 206, Figure 3)

Table 3.2 The urban population of England and Wales 1801–1911 (after C. M. Law 1967, 130)

	Urban population	Urban population as per cent of total	Decadal change per cent	Urban change as per cent of total change
1801	3,009,260	33.8		
1801–1811			14.0	56.0
1811	3,722,025	36.6		
1811–21			18.1	59.0
1821	4,804,534	40.0		
1821–31			15.8	71.1
1831	6,153,230	44.3		
1831–41			14.3	76.3
1841	7,693,126	48.3		
1841–51			12.6	94.1
1851	9,687,927	54.0		
1851–61			11.9	98.0
1861	11,784,056	58.7		
1861–71			13.2	151.8
1871	14,802,100	65.2		
1871–81			14.7	103.5
1881	18,180,117	70.0		
1881–91			11.6	113.0
1891	21,601,012	74.5		
1891–1901			12.2	107.0
1901	25,371,849	78.0		
1901–1911			10.9	87.4
1911	36,070,492	78.9		

1891. But perhaps the figures given by Waller (Waller, 1983, 25) are nearer to the actual situation, for whereas the population of the London County area increased from 2.81 millions in 1861 to 4.54 millions in 1901, the population of the outer ring increased from 0.41 million to 3.0 millions so that the Greater London conurbation increased from 3.22 millions to 7.54 millions in the second half of the century. Thus London increased its share of the country's total population from some 9.7 per cent in 1801 to either 14.6 per cent or 19.5 per cent in 1901, dependent on the definition adopted. If the earlier years of the century were dominated by the population increases in the new large industrial cities, after 1881 London's increase was the predominant characteristic.

It is possible to take the argument further and suggest that the whole industrialization and urbanization process was a consequence of London's role. E. A. Wrigley has set out the case in the form of a diagram (Figure 3.6) which indicates the links between London's growth and the industrial revolution (Wrigley, 1978). This simply demonstrates the way in which demands set up by London's size, which greatly exceeded that of any other city, triggered off those changes which are now regarded as critical in the industrial revolution. In the diagram, the central boxes between London and the precipitated changes in the provinces can be interpreted as having critical influences on the regional sub-systems. Improved transport, better commercial facilities, higher real incomes, agricultural change, all these were stimuli which promoted the growth of provincial towns as they permeated the country.

Such an interpretation would provide a neat relationship with the notion of the growth of a world system and relate national events to a broader scale of change. The centre of political and economic power lodged in the Mediterranean in Roman times had swung under the Holy Roman Empire to a continental and land-based system. It was the growth of a much wider ranging maritime trade which undermined that condition and led to the growth of those political entities fringing the Atlantic. But Spain, the inheritor of the continental power but with an Atlantic coast, was too tied to the past and it was the powers to the north which experienced uninhibited growth. Free from the smothering power of the nobility, the urban bourgoisie were able to develop significant mercantile operations which gave them the necessary political clout and accumulations of capital to invest and engender the effective change to a factory-based, capitalist system. London marked the apex of mercantile endeavour which triggered all those 10 aspects of change which are entered in Wrigley's diagram (Figure 3.6). Subsequently, the import of raw materials and the export of finished goods sucked colonial or dependent territories into the system, setting off an urbanization process there too. But that is a different story.

The nature of the growth of London's population set the pattern for the rest of the country. It is an assumption generally made that the great conurbations of the nineteenth century were the products of migration. That is undoubtedly true, for after all, it was the significance of such movement which produced an academic 'spin off' in Ravenstein's *Laws of migration* (Ravenstein, 1885). Figure 3.5, which shows net migration currents, that is the balance between out- and in-migration, is derived from the 1861 census evidence as to place of birth. As an aside it is worth noting that the author of the map indicates that it is the first census year for which such a map can be drawn. Flows into London dominate the map but there is also an obvious distance-decay relationship with movement to London becoming less significant with distance from London. Indeed, it is feasible to contend that the map displays a north–south divide with London completely dominant south of a line from the Severn to the Wash, but of considerably less importance to the north of it.

Two characteristics of the immigrant population need to be noted. The first was

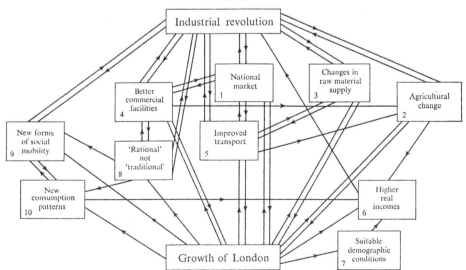

Figure 3.6 Links between London's growth and the Industrial Revolution in England and Wales.
(after E. A. Wrigley, 1978, 240)

that, as Ravenstein promulgated, women were more migratory than men so that sex ratios were biased towards women. The reasons were the obvious ones, the decline in agricultural work with an increasingly capitalized agriculture, the burgeoning of domestic service in the city and the clearly enhanced chances of marriage. The other characteristic was that migration was also age-specific. The majority of immigrants were young with the 15–30 age group constituting the largest contributor. 'Statistics presented to the Royal Commission on the Poor Law (1905–9) showed that in London old people over 60 constituted 66.9 per 1,000 population. In other wholly or mainly urban districts they formed 67.3 per 1,000, in wholly or mainly rural districts 102 per 1,000' (Waller, 1983, 26).

As the century progressed the congregation of such young age groups in the capital, with a preponderance of marriageable females, produced a high birth rate so that by mid century the major contribution to population growth was not migration but natural increase by excess of births over deaths, that is in spite of the overcrowding and lack of sanitation which produced high death rates. Thus, as Waller states, between 1851 and 1891 84.03 per cent of the increases in London's population resulted from the surplus of births, 15.97 per cent from immigration. Between 1851 and 1911, of the 170 per cent increase in London's population from 2.7 millions to 4.6 millions, 134.2 per cent was accounted for by natural increase and only 35.8 per cent by migration.

If Figure 3.4 be examined, however, and more especially the detailed inset of London, another trend becomes apparent. By the end of the century central London was losing population. Outmigration from the centre to the growing suburbs was taking place, and a ring of areas where net immigration exceeded natural gain is visible. Indeed, at the centre natural loss was apparent. 'The one square mile of the City had a resident "sleeping" population which peaked around mid-nineteenth century at just short of 130,000. This had dropped to 112,000 by 1861; then it fell to 51,000 by 1881, 27,000 by 1901 . . .' (Waller, 1983, 28). This is reflected in the figures for the Greater London group of 'residential towns' which show a total growth of 191 per cent between 1851 and 1911, where 108 per cent is ascribed to natural increase and 82.9 per cent to migration, compared with the 35.8 per cent for the County of London.

Urban population growth: natural increase and migration

This brief treatment of London sets the pattern for the rest of the country, where each regional system as it developed showed an initial dominance of immigration followed by the greatest phase of increase when natural gain predominated. But it was generally a highly complex period when high death rates offset high birth rates to give immigration a highly variable significance. By the late century both the onset of industrial decline and, in the thriving areas, the impact of surburbanization produced declining city populations.

The pivotal role ascribed to London by Wrigley, however, is open to challenge. The actual process seems to have been one of the emergence of a number of separate regional areas of growth stimulated by parallel factors. These areas eventually produced a local coherence in the interrelationship of centres and being merged into a national system, thereby giving London and its region, the south-east, as the capital of that system, the predominant role which it increasingly occupied in the nineteenth century. But this anticipates the formation of the urban system to be treated in the next chapter.

Two further examples can be added to the brief review of London's pattern of growth. Peter Large in a paper on 'Urban growth and agricultural change in the West Midlands during the seventeenth and eighteenth centuries' has demonstrated how the cluster of towns in the area came into being with Birmingham as the capital. He points out that metallurgy had been important since the middle of the sixteenth century, largely stimulated by the demand from the farming countryside for products such as buckles, spurs, stirrups and bridle bits, as well as for nails and tools such as scythes. It was mainly located in South Staffordshire. Expansion followed on this basis. 'The hardware trades were primarily orientated towards the domestic market and supplied commodities in constant agricultural and domestic use to the regions of Midland, Southern and Western England. During the final quarter of the seventeenth century the growing export trade to the American and West Indian colonies and to Europe supplemented the domestic markets but did not supplant them' (Large, 1985, 170). The response was a rapid growth in population. Thus, in Sedgley, a centre for nail making, the population had increased from 600 in 1563 to 2,300 in 1663. (Frost, 1973, 283–376). Supremacy accrued to the marketing centres dealing in bar iron and the finished products, especially Wolverhampton and Birmingham. Birmingham itself became the prime commercial centre with an increasingly sophisticated manufacture of guns and of buckles and buttons, from which its jewellery trade was to develop, and of objects in brass.

By the first census of 1801 the population of the Birmingham Registration District was 60,822, but that excluded Aston with a further 15,310. By mid century the population had risen to 173,951 with 66,852 in Aston. By the last census of the century in 1891 the population of the extended County Borough was 478,113, the comparable area having been 70,670 in 1801 and 232,841 in 1851. By 1891 although there was an excess of births over deaths of 28,773 in the Registration District for the previous decade the actual population had fallen by 850. Indeed, between 1861 and 1871 a similar natural increase of 28,727 had been recorded but with an actual increase of only 18,394. Thus, for most of the second half of the century outmigration characterized the city as the development of suburban areas became more significant. Birmingham is representative of the way in which throughout the country separate regional agglomerations of population took place, at least to a considerable degree initially, independent of London's influence. It was as small individual entrepreneurs gave way to the large amalgamated firms that control moved more effectively to London. But rather more detail as to the actual population changes can be derived from the second example.

The second example is Liverpool which has been extensively investigated by Lawton. As he writes, 'the parish registers of the late eighteenth and early nineteenth centuries reveal one of the most explosive phases of growth in any modern British city. In the 1790s the population of Liverpool grew by 40 per cent (some 22,000) perhaps two-thirds of which was due to migration. From 77,653 in 1801 the Borough's population increased to 286,487 in 1841, due principally to rapid in-migration' (Lawton, 1983, 218–9). In consequence of course, there was a high birth rate but not a high rate of natural increase since Liverpool, owing to some of the most insanitary and overcrowded conditions in England, continued to experience a very high mortality, especially following the Irish in-migrations of the 1840s and the typhus epidemics of that decade. The death rate reached as high as 71 per 1,000. As a result Liverpool showed a greater increase due to migration rather than natural increase well into the second half of the nineteenth century. 'Thus the rapid growth of population to 493,405 in 1871 owed more to migration than to natural growth, a migration gain of 211,685 in Liverpool's three registration districts as compared with 96,992 natural increase' (Lawton, 1983, 219). After 1871 the still high natural increase of the population, and declining death rates, took over, so that the period 1871–1911 showed a natural increase of 746,421 as against a migration loss as the outward suburban movement began.

These brief analyses of Birmingham and Liverpool suggest the existence of a crossover point (Keyfitz, 1980, 142–56) where natural increase becomes greater than migration. In Lawton's tabulation of registration districts in the urban areas of England 46.6 per cent had achieved that point before 1851, with a further 12 per cent by 1861 and 12.6 per cent in the decade 1861–71 so that by 1871 some 71 per cent of the urban registration districts had experienced the crossover to natural increase dominance. This is confirmed by Figure 3.4 where the category 'net in-migration exceeds natural gain' is greatly restricted and characterizes mostly the late-developing residential suburban and seaside areas.

This issue of the significance of natural increase, of the demographic transition as against the mobility revolution, has been extensively reviewed by de Vries (1984), but with evidence largely derived from continental Europe. He argues that his analysis 'will require a modification of the conventional view that nineteenth-century urbanization was primarily fed by an unprecedented increase in migration constituting a "mobility revolution"' (de Vries, 1984, 234). But as this chapter has indicated, early high migration rates were followed by a predominant natural increase in British cities. But de Vries's broader conclusion, 'that the rapid urban growth of the nineteenth century need not have been accompanied by a major change in migration flows' (de Vries, 1984, 233) seems unacceptable in relation to England and Wales. That is equally true of his view that 'the quickened pace of urban growth in the nineteenth century did not so much signal an increase in rural–urban migration as it reflected a change in the internal demography of the European city' (de Vries, 1984, 233).

The spread of urbanization

Friedlander's map (Figure 3.3) of the progress of urbanization generalizes these trends although its crude county basis limits the detail and distorts the true pattern. Outside London, Lancashire and Durham are the earliest counties to be classed as urban, followed by Glamorgan, Staffordshire and Warwickshire. 'Thus by 1871, urbanization . . . was concentrated in four areas, Durham and Lancashire in the North, Staffordshire and Warwickshire in the Midlands, the two South Wales coal counties and around London. Already in that year some 38 per cent of the total population lived in these urbanized counties . . . However, during 1871–81 urbanization spread from the northern counties southwards at increasing speed . . . by 1891 the whole northern urban area from Northumberland down to Yorkshire had joined with the three Midland urban areas (Staffordshire, Warwickshire and Derbyshire) to form one contiguous urban area' (Friedlander, 1970, 430). That is an overgeneralized and bland view of a process much more regionally complex, but it epitomizes the processes of urbanization at work.

Those processes can be more directly associated with 'the differential growth of urbanization in distinctive phases of development' (Shaw, 1989, 57), the phases related to Kondratieff waves. Thus, the first phase from the 1770s to 1820 saw the 'industrial revolution Kondratieff' (Table 3.3) where cotton textiles based on steam power together with the iron industry were the critical elements. The period 1842–97 was the 'bourgeois Kondratieff' where the impact of the railways spread development wider and the coal–steel–engineering complex dominated accounting for Friedlander's spread of urbanization. The third Kondratieff after 1897, the

Table 3.3 Kuznets's Scheme of Kondratieff Cycles and Mensch's Innovation Phases (after Shaw, 1989, 58)

Economic characteristics	Kondratieff Cycles			
	Prosperity	Recession	Depression	Revival
1787–1842. 'Industrial Revolution Kondratieff': cotton textiles; iron; steam power.	1787–1800	1801–1813	1814–1827	1828–1842
1842–1897: 'Bourgeois Kondratieff'; railways; steel; engineering; coal.	1843–1857	1858–1869	1870–1885	1886–1897
1897 to date: 'Neo-mercantilist Kondratieff': electricity; oil; chemicals; cars; light engineering.	1898–1911	1912–1925	1926–1939	

neo-mercantile wave, extended urban growth across the whole country, but essentially switched to the south-east accompanied by the beginning of the change of emphasis from the heaviest industries to the lighter engineering trades and to the tertiary and quaternary sectors.

Conclusion

At this point it is appropriate to return to the more general scale adopted at the beginning of the chapter. The period under review, an arbitrarily extracted 'nineteenth century' saw the rapid development of the space economy in the form of a series of separate industrialized and urbanized regions. The roots go much further back, even to a proto-industrial phase, but the key to the specific urban transformation was the factory system. This was introduced as capital sought the more profitable areas, economically and locationally, in which to invest. Elsewhere, as in the coal mines, there was no choice other than larger and larger-scale operations. Factory and mine demanded labour which, given the pre-industrial distribution of population, was not locally available so that inmigration dominated the early phases of urban growth. But, inevitably, and in spite of all the sanitary and health problems of the cities, natural reproduction took over and natural increase dominated, as set out in the models with which the chapter began; the demographic transition was under way. This produced the massive conurbations which Geddes was to identify in 1917 (Geddes, 1917) as the significant creations of the nineteenth century. But it also produced a wide array of towns, that 'width' referring as much to function as to size. By the end of the century however, falling fertility and the recognition of the diseconomies of scale had led to declining city-centre populations and the initiation of a period of suburban sprawl which was to dominate England and Wales until the Second World War.

4

The changing structure
of the city system

The significance of the general process of urbanization in England and Wales, of the growth of the urban population, has been discussed in the last chapter. The parallel but somewhat different topic of the nature of the system of cities created by that process was there set aside to be considered in this chapter.

The issues which need to be discussed are perhaps best put in a series of questions. They will not be answered separately and seriatim in this chapter, but they establish the major problems which arise.

1 Was there a revolutionary overturning, a total restructuring, of the city system? What was the degree of continuity between the pre-industrial and the industrial city systems?
2 To what extent did agglomeration economies and the hierarchical introduction and spread of innovations play a significant role in structuring the city system?
3 Was there at the outset, at the beginning of the industrial period rather than necessarily at the beginning of the century, an integrated system of cities organized on an identifiable principle, through which the waves of change, brought about by industrial growth, ran? Or was there a series of separate regional systems which, during the century, were progressively organized into a single national system? If the latter is the case, when can that single national system be identified?
4 What was the impact of the great widening in the functional specialization of towns after 1750 and what role did it play in determining the character of the system?
5 What was the nature of the city system itself? Did it display evidence of primacy, of a hierarchical ordering or of a rank–size relationship? If orderings can be identified did they change during the century, and if so, why?

These five questions do not cover all the issues, especially those of local detail, and they are not mutually exclusive for they overlap and, indeed, run together as the system of cities is considered. Even so, they do represent the major discussion points which a review of the nineteenth century engenders.

Transformation or continuity in the city system?

The first of the questions as to the exact nature of the transformation which took place or, indeed if there was any such transformation at all, touches on the most fundamental issue. As there has been a fashion among economic historians to write off nineteenth-century industrialization as constituting anything which merits the epithet 'revolution', so too historical geographers and historians have tended to argue against any revolutionary restructuring of the city system during the same period.

The most thorough analysis of this question and the most cogent arguments for nineteenth-century continuity as against revolutionary transformation have been put by Jan de Vries, briefly in a chapter entitled 'Patterns of urbanization in pre-industrial Europe' in *Patterns of European Urbanization from 1500* edited by H. Schmal (Schmal, 1981) and more extensively in his own book *European Urbanization 1500–1800* (de Vries, 1984). Undoubtedly some of the difficulty in using De Vries's work in the context of this chapter is that it is based on the broadest scale of analysis, for it is European in coverage, and it is one of the axioms of geographical interpretation that explanation changes with scale changes. Even so, much of de Vries's argument relates to the evidence of individual countries. In briefest summary, a mode of presentation which does nothing like justice to the vast array of data or urban population meticulously assembled, de Vries contends that until the sixteenth century the cities of Europe were characterized by a polynuclear system, that is, there was a series of semi-independent and separate urban systems across the continent. Subsequently these were integrated into a single system and that occurred with the rise of North West Europe to dominance. The decisive period of that transformation was the first half of the seventeenth century when the Atlantic and North Sea coasts of Europe became the controllers of trade. The Dutch East India Company was founded in 1602, the West India Company in 1621, the English East India Company in 1660, the English Hudson Bay Company in 1670 and the Royal African Company in 1672. These were the bases of a new and dominant mercantile regime. It will be apparent that there are parallels with the work of Braudel and Wallerstein for what de Vries is presenting is the specific urban aspect of a world system. The critical discontinuity occurs, therefore, on a world scale, not in the nineteenth century but much earlier when the hegemony of the Mediterranean was finally set aside. Spain, the power with locational links with both the old and the new had led the way but failed to respond to the new challenges because of its social sclerosis. 'By 1686 the surge in overseas trade had pushed the total (of English shipping) to 340,000 tons. After a temporary decline in tonnage around 1700, the momentum of growth recovered with tonnage reaching 1,055,000 by 1788, a twenty-fold increase since the late sixteenth century. Even before the end of the sixteenth century, the size of the English merchant fleet and the value of the trade being handled surpassed that of the Dutch, and English hegemony over international trade had started to replace Dutch hegemony' (Dodgshon, 1987, 294–5).

These hegemonies via the cities they were based upon meant a complete reorientation of urban growth from which a transformed system of cities developed, with the economic clout and the transport linkages to generate an integrated, single European system, indeed, a world system. Nineteenth-century industrialism simply grafted greater growth onto this existing system. 'The great discontinuity must be sought in the seventeenth century when Europe's polynuclear urban system was made to give way to the leadership of one of its centres, that of north-western Europe' (de Vries, 1984, 172). Again de Vries writes 'the medieval city may have had many virtues, but the autarchic urban structure of the Middle Ages could not serve as the urban framework of a commercial and industrial society. The industrial city of the factory age was undoubtedly a novel and powerful organism, but it inserted itself into an existing urban system which it

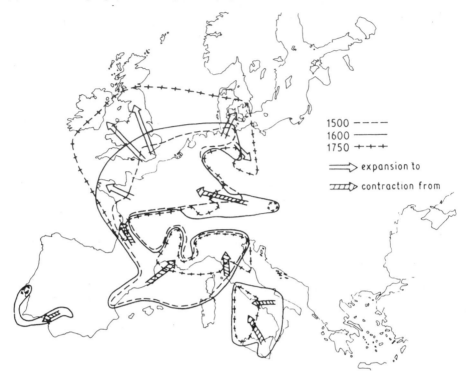

Figure 4.1 Zones of expanding and contracting urban potential: 1500, 1600, and 1700. (after de Vries, 1984, 166, Figure 8.8)

modified but did not transform' (de Vries, 1984, 254). In passing, it must be noted that de Vries's rejection of immigration as the significant feature in nineteenth-century urban growth and his criticism of a mobility revolution (see page 40) are at one with his scepticism as to a nineteenth-century transformation of the city system. De Vries's general interpretation is summarized in Figure 4.1 where zones of expanding and contracting urban potential are shown at three dates, 1500, 1600 and 1750. Urban potential is derived from the more standard concept of population potential, the two measured elements being city size and distance. De Vries, however, refines from the crude measures to give more acceptable bases. The map demonstrates the passage of dominance to the north-west, which by the eighteenth century was integrating the whole European system into one.

At this point it is appropriate to turn to the second question posed at the outset of this chapter and to introduce the concept of agglomeration economies and cyclical growth. The geographer who has written most extensively on this subject is Alan Pred, who has published a large corpus of work which is both theoretical and empirical in its content, although the latter aspect is concerned exclusively with the United States of America.

In order to present the core of Pred's work two diagrams are reproduced (Figures 4.2 and 4.3). Again this brief rehearsal and arbitrary extraction of illustrations does no justice to the richness of the originals. The first diagram (4.2) outlines the circular and cumulative feedback process for a single large mercantile city between 1790 and 1840. The second (4.3) epitomizes the parallel situation for individual large cities during initial modern industrialization. Clearly, they both depict generally similar processes. In the earlier mercantile period the expansion of the trading and wholesaling complex generated not only local earnings which fed back directly into further growth, but stimulated an increased scale of operations and increased the possibility of invention and innovation, which in turn stimulated the enlargement of industry. The population increase brought about set up new and larger local and regional thresholds so that new and larger-scale industries were attracted. The result was a cyclical process by which the greatest growth was generated by the largest cities – to those that had was given.

This same process in the context of large cities in the later industrial period is demonstrated in Figure 4.3 where, similarly, new or enlarged industry, located in the large mercantile cities by all that is signified by Figure 4.2, had the same impact. Infrastructural enhancements act as attracters of economic activity contributing also to the introduction of new methods and products, invention and innovation. Thus the cyclical pattern of growth is sustained.

In Pred's work there would seem to be a considerable element of support for de Vries's theme. In Europe, the location of mercantile activity along the coastlands of the north-west was the critical period of change and subsequently, the two phases illustrated in Figures 4.2 and 4.3 maintained the pre-eminence of the early cities.

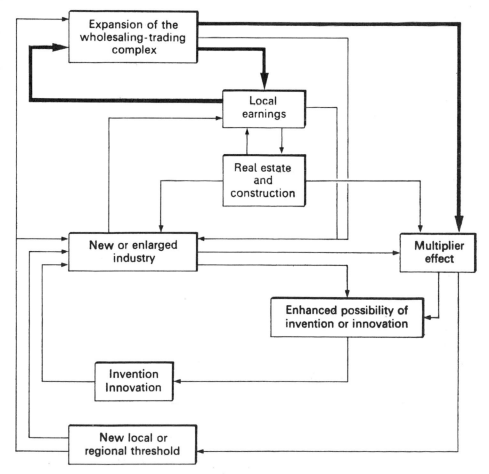

Figure 4.2 The circular and cumulative feedback process of local urban size growth for a single large mercantile city.
(after Pred, 1977, 72, Figure 2.8)

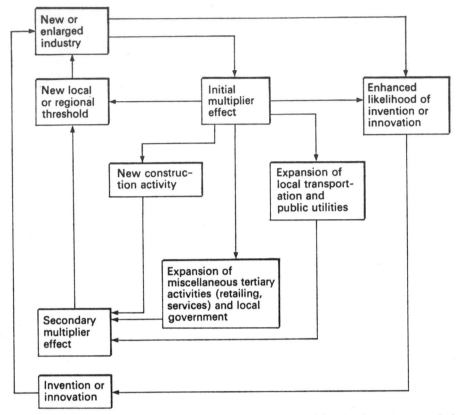

Figure 4.3 The circular and cumulative feedback process of local urban size-growth for individual large cities during periods of initial modern industrialization. (after Pred, 1977, 90, Figure 2.16)

Mercantile increases led to the stimulation of industry so that industrial magnitude succeeded to mercantile significance and the urban hierarchy remained stable and unchanged. Thus Pred writes of 'the concurrent spatial concentration of manufacturing and maintenance of large-city rank stability in the US between 1860 and 1914' (Pred, 1977, 94). The critical characteristic is rank stability and although the dates in the quotation are very late, nevertheless the general interpretation is that of de Vries, that the nineteenth century, or rather more extensively the industrial period, saw no major changes in city rank. There was a great deal of jostling and minor change among the lower-order towns where differentiation was in any case minimal. At the higher levels of the hierarchy there were none or few changes.

   This examination of the more general or theoretical notions, and taking into account the empirical work of de Vries and Pred, seems to have provided an indicator for the interpretation of nineteenth-century Britain, and answers the first two questions set out at the beginning of the chapter. There was, it is implied, no transformation but rather major changes took place much earlier in the seventeenth century, and the operation of all those forces which can be encapsulated in the concept of agglomeration economies confirmed and indeed accentuated the dominance of the existing large cities. True there was very considerable movement or rank changes at the lower levels but they were local and of no great moment. The greatest example of continuity was London with its influence and growth as they were set out in the diagram by Wrigley (Figure 3.6) in the last chapter, which can well be considered as a particularization of the generalized diagrams of Pred.

Changes in the rank order of towns

It is now appropriate to move to a more empirical review of the actual changes which took place in the rank order of towns in England and Wales during the industrial period, roughly beginning near the middle of the eighteenth century. The conclusions can be related to the more general discussion with which the chapter began. Two pieces of evidence can be introduced. The first, Figure 4.4, is a diagram of changes of rank order in the 25 largest towns of England and Wales between 1801 and 1911 constructed by Robson (Robson, 1973, 39) using Law's data (Law, 1967). The second, Table 4.1, is a listing of the higher order towns of England and Wales at dates between 1500 and 1800.

The evidence from Robson's diagram seems to support de Vries's contention. The graph can virtually be divided into three sections. At the highest level there were barely any changes at all over the century. Below these distinctive leaders there are changes but also evidence of stability, whereas in the lowest section there were quite considerable changes. Even so, there are some reservations which must arise for the diagram deals only with the twenty-five largest towns and begins at a time when industrialization had already been developing for over half a century.

If one turns to a longer time span then a somewhat different conclusion can be drawn with quite significant changes having taken place even at the highest levels. Thus between 1650 and 1801 of the first seven towns at the earlier date only two,

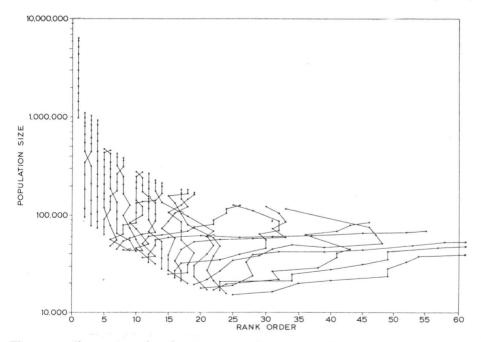

Figure 4.4 Changes in rank order 1801–1911. Successive ranks of the 25 largest towns in England and Wales in 1801.
(after Robson, 1973, 39, Figure 2.5)
Note. Bristol, which was ranked 5 in 1801, seems to have been omitted from this diagram.

Table 4.1 The estimated populations in thousands of English towns between 1500 and 1800 (after de Vries)

| | Populations in | | | | | | |
	1500	1550	1600	1650	1700	1750	1800
Bath	0	0	0	1	3	9	32
Birmingham	0	1	2	4	7	24	69
Blackburn	0	0	0	0	0	4	12
Bolton	0	0	0	0	0	4	13
Bristol	10	10	11	20	25	45	64
Cambridge	0	0	0	9	9	6	10
Carlisle	0	0	0	0	4	4	10
Chatham	0	0	1	0	5	5	11
Chester	0	0	0	8	10	13	15
Colchester	5	0	0	10	10	9	12
Coventry	7	7	7	7	7	12	16
Derby	0	0	0	0	0	6	11
Exeter	10	8	10	10	14	16	17
Greenwich	0	0	0	0	0	UNK	14
Huddersfield	0	0	0	0	0	0	11
Hull	0	0	0	0	6	6	28
Ipswich	0	0	0	UNK	8	12	11
King's Lynn	5	0	0	5	5	9	10
Leeds	0	0	0	5	6	16	53
Leicester	4	3	5	5	6	8	17
Liverpool	0	0	0	0	6	22	78
London	40	80	200	400	575	675	865
Manchester	0	0	0	5	9	18	70
Newcastle	10	10	10	13	14	29	28
Norwich	10	12	12	20	29	36	37
Nottingham	0	3	5	6	7	12	29
Oldham	0	0	0	0	0	0	12
Oxford	5	0	0	9	8	8	12
Plymouth	2	4	7	7	9	15	43
Portsmouth	0	0	0	4	5	10	32
Preston	0	0	0	0	0	6	12
Reading	0	0	0	0	0	7	10
Salford	0	0	0	0	0	0	14
Sheffield	0	0	2	0	10	12	46
Shrewsbury	5	0	0	0	10	13	15
Stockport	0	0	0	0	0	3	15
Sunderland	0	0	0	0	5	10	24
Warrington	0	0	0	0	0	4	11
Wenlock	0	0	0	0	0	UNK	15
Wigan	0	0	0	0	0	4	11
Wolverhampton	0	2	0	4	0	7	13
Worcester	0	3	6	8	9	10	11
Yarmouth	5	0	0	10	10	10	15
York	8	8	12	12	11	11	16

0   = Population below 10,000
1–9 = Population below 10,000 but given for information only
UNK = Unknown

one of which was London, appear at the later, suggesting considerable reordering after 1650. Even de Vries's own tabulations show that the total number of towns with populations of over 10,000 increased at the four dates of 1600, 1650, 1700, 1800, with 2, 3, 10 and 23 respectively. Any crucial change seems therefore to have taken place after 1700 and particularly after 1750.

Any effective discussion of the longer time span necessarily carries investigation back into the eighteenth century. The urban system at that century's start was that which had emerged from the late medieval period. It was controlled by local wealth and population numbers, but it also reflected political influence through the administrative system so that those towns which had risen out of the generally undifferentiated lower level owed a great deal to the administrative functions, both lay and ecclesiastical. Towns which had had counties named after them as an indication of dominated areas, both politically and economically, still retained their higher status. Such, for example, were Nottingham, Derby, Leicester, Northampton and Bedford. Above these were the provincial capitals and there is agreement that five stand out – Norwich, Bristol, York, Exeter and Newcastle. All had maintained their status over some two hundred years. They had a number of features in common. They were the focal points for wealthy tributary areas and were located on navigable rivers or functioned as ports. All had significant political, administrative and ecclesiastical roles. None was within competing distance of London. Also, to a degree, these cities were the focal points for the mercantile period. An even earlier period of change had seen the transfer of significance to ports of trade, as for example, the growth of Southampton as capital of the Hampshire basin at the expense of Winchester. These, then, were the real centres of mercantilism. But they were not to become the prime centres of industry. By 1801, if London be excluded, none had retained its rank; Bristol had fallen to fifth place, Norwich to seventh, Newcastle to thirteenth, Exeter to fourteenth and York to fifteenth. By 1851, although Bristol remained at fifth and Newcastle had risen to tenth, Norwich had fallen to seventeenth and Exeter and York were not among the first twenty. The rank stability apparent in Robson's diagram is, therefore, a product of the date of the first census and although strictly the nineteenth century did see few changes at the highest levels, the industrial period did result in fundamental changes, to a degree, far more significant that the slotting into, or the filling out of, an existing system. But if this were the case then one needs to consider the role of innovations to which considerable significance has been allotted.

One of the keys to what must seem the uniqueness of England and Wales, which de Vries partly accepts (1984, 171), is the fact that they were the first countries to be industrialized and consequently experienced that transformation before the establishment of a new rail-based communication system. The opposite was the case in most European countries. It is a universal phenomenon that priority in communications development and improvement is given to links between the largest cities, as a logical response to demand. The result, however, is that such favoured cities grow even more rapidly, generating further demands on communication development. Thus favoured cities and regions grow excessively, the core advances and the periphery becomes even more neglected and stagnant. Such is the link between innovation and agglomeration economies. But in England and Wales in the late eighteenth century industry grew in relation to specific locational advantages and the early transport system was based on industrial demand, industry created the transport network and in doing so at least in part created the system of cities.

Some more general reasons can be derived by following part of the argument set out by Dodgshon in his treatment of spatial change during the industrial revolution in his book *The European Past* (Dodgshon, 1987). In essence a conflict arises between investment in fixed capital, which was increasingly significant with factory-based industry, and free or circulating capital. As fixed capital becomes less usable or adaptable and the freedom to adapt becomes exhausted, capital shifts to where it can most profitably be invested. 'The value of what is invested in fixed capital is written off or, in reality, transferred as a burden to others' (Dodgshon, 1987, 323). This is especially true in the socio-economic sphere where areas having provided profit are simply abandoned leaving the community, in modern jargon, 'to pick up the tab'. Reinvestment takes place in the new areas offering maximum returns. 'Construed in this way, the history of capitalism should read as a series of locally – or regionally – specific investments, each moving through a cycle of innovation, adaptation, then crisis as capital moves on to areas of unused freedom' (Dodgshon, 1984, 324).

Now the critical phrase in the quotation above is 'unused freedom'. Investment is quite deliberately avoiding not only the outmoded investment of early areas, it is also moving away from all those constraints which could be imposed by an established communal structure. It moves to new areas and there creates new towns.

Dodgshon's interpretation offers two further indications relevant to the urban system. The first is that the concept of industrialization developing through, what he calls a series of place-specific adaptations, is generally in agreement with much that has been written about the industrial revolution. He quotes Pollard (1981) and Langton (1972) as scholars who have stressed the fact that industrialization was a regional process. Indeed, Langton views attempts at interpreting industrialization through national aggregates as fruitless since the vital dynamics of change were those which combined around specific locations and regions (Langton, 1972, 54). The implication for the urban system is that a whole series of regional sub-systems evolved, each related to a specific locality and only very partially integrated into the national aggregate, the traditional rank–size graph which urban geographers prepare. But there is no anomaly between the lack of national integration suggested by one approach and the apparent effective integration shown by the other, for the approximation to the rank–size rule can be related to the amalgamation of a series of semi-independent systems, indeed it has become the basic explanation of its deviation from a hierarchical ordering.

At this point Robson's study of the impact of functional differentiation within systems of industrialized cities can be introduced (Robson, 1981). His main theme derives from the rather nebulous distinction between basic and non-basic, or city-forming and city-servicing functions (Carter, 1981, 45). Basic or city-forming functions tend to be highly specialized and because of their specialization and dependence upon a very narrow functional base, they are also the most variable in growth; they are also the most vulnerable to economic change – 'this measure of growth variability does provide some support for the extreme form of the hypothesis; places which are most highly variable are indeed those which are specialized, highly dependent upon a limited range of activities, most highly functionally differentiated and therefore most critically dependent upon the circulation and shift of the social surplus within the economy' (Robson, 1981, 127). So, high growth carries risk of great fluctuation in fortune; the mushroom is an analogy greatly used by urban historians and geographers. His argument, Robson contends, is based on the concept of growth as a function of the circulation of the surplus. This is quite clearly at one with the interpretation by Dodgshon which has been outlined and confirms the conclusion as to essentially regional and fluctuating patterns of growth.

But Robson also notes the corollary, that widely based towns were those that experienced more steady growth. The town which begins as a one-industry town initially lacking in service provision will, if it can attract further investment, grow and develop service functions. 'At this extreme, such "manufacturing" centres as Newcastle and Manchester develop as organizing nodes for a regional system of manufacturing activity and, with this, comes the development not only of the financial enterprises which lubricate the regional productive activity, but also retail, warehousing and commercial activities which are non-industry specific' (Robson, 1981, 120). This is much nearer, of course, to the interpretation of de Vries and Pred, for these towns were often the early centres of what can be called proto-industry. The example of Birmingham was briefly considered in the last chapter. But, even so, they are the creations of industry. They are also the towns which Pred would cite as the examples of cyclical growth, the towns already the largest in 1801 which would be the largest in 1901 and indeed, in all probability, in 2001.

The second comment by Dodgshon relevant to the city system now fits into the theme being developed, 'We must not overlook the fact that, increasingly, as capital markets became organized, eruptions of regional or local activity need to be seen as overlain by a higher, more national and ultimately international circuit of capital that sought to engage and disengage from such place-specific investment, as new breakthrough opportunities offering higher profit rates presented themselves (Dodgshon, 1987, 325). In city-system terms this reiterates Robson's identification of the broad-based service settlements which developed as the centres of financial manipulation and administrative control. And at the top was London. Moreover, in parallel with the sort of evolution of the city system throughout the century was the evolution of business. This is a complex issue and apparently marginal to the city-system development but in fact it is in many ways central, for the period when entrepreneurs became dependent on the London financial markets was the point at which London effectively controlled the whole of England and Wales. The century, and certainly industrialization, had begun with small-scale entrepreneurs deriving their capital from local sources. It is possible to suggest a progression throughout the country beginning with the predominance of the local owner-manager. 'The century between 1750 and 1850, before the invasion of industry by joint-stock finance and limited liability, was the heroic age of the private owner of risk capital' (Court, 1954, 87). Partnerships were the predominant means of development and of increase in scale. But with the increase of scale a separation began between the entrepreneur as financier combined with the capital market and the on-the-spot managers of businesses. This eventually developed into a division between a series of specialist managers who retained the link with locality and entrepreneurs closely connected with the trading and financial markets making the strategic decisions and more and more metropolitan in their orientation. Clearly the Joint Stock Acts of 1856 and 1862 were crucial in this process. Financial services, too, moved in the same direction, the multiplicity of local banks at the beginning of the century giving way to the national amalgamations by its end. This process should not be overstressed. P. L. Payne in his book, *British entrepreneurship in the nineteenth century* (1988) argued that there is little evidence of significant divorce of control from local ownership before the end of the century and that the dominance of the private company only gave way to the large-scale joint stock company in the decades before the First World War. Such a view would not be at odds with the suggestion in this chapter that the complete integration of the city system did not come about until very late in the period under review and indeed, not until the beginning of the twentieth century.

Similar evidence can be derived from retailing, which in the first flush of central-place studies was commonly used as the basis for the ranking of towns. Thus Jeffreys demonstrated that the grocery or provision chains with 10 or more branches numbered 31 in 1885, those with over 25 numbered eight. By 1910 these figures had risen to 114 with 10 or more branches, 44 with 25 or more (Jeffreys, 1954, 137). Again, as an example, Jesse Boot established the business in proprietary medicines in Nottingham in 1874. By 1900 he had 181 branches; by 1914 that figure had increased to 560. Only by the latter date is national penetration sufficient for the shop itself to be used as a measure of urban rank, that is when all parts had been assessed for potential colonization.

The emergence of a system of cities

A point has now been reached where it is possible to suggest an answer to the third question posed at the outset of this chapter, that as to the period at which an

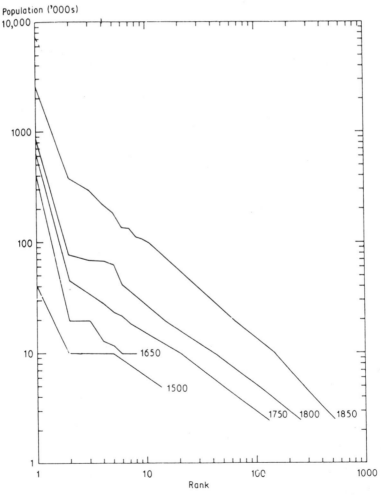

Figure 4.5 Rank-size distributions in England and Wales 1500–1974. (after de Vries, 1984, 119, Figure 6.18)

integrated city system can be identified. The impact of industrialization was not only to send a ripple through an extant urban system, although that process certainly occurred. Robson in *Urban Growth* demonstrated effectively the way innovations in the form of the spread of gas works, at the start of the century, building societies in the middle years and of the telephone in the final quarter, illustrated the associated processes of hierarchical diffusion down the ranks of the urban size array and thence outwards in a local neighbourhood effect (Robson, 1973, 184). Probably the same could be shown for the spread of Boot's chemist shops. But whereas that ripple is significant in converting the individual complexity of retail and service arrays from the local diversity of the century's beginning to the standardized 'trait complexes' of the century's end, of critical importance were the organizational structures of enterprises and undertakings which played their part in creating a single integrated urban system in England and Wales. In directly operational terms, or the procedure of academic analyses, there comes a critical point when a switch can be made from the presence/absence of trading types or shop types to the identification of complexes of named chain stores, banks, building societies and other indicators of urban status. It is at that point that a unified system is assumed to exist and though it has never been identified, no study of the period before 1914 has used the latter approach.

Another aspect of urban development must be introduced at this juncture. As already indicated (p. 50) the nineteenth century saw a great elaboration in the special functions of cities. Even the Census was forced into presenting a classification of urban settlements in its analysis of population growth. Initially introduced in 1851, that classification became progressively more elaborate, as Table 4.2

Table 4.2 The Census classification of the towns of England and Wales by function in 1861 (Appendix, Table 52)

Class	Numbers in each class
1 London	1
2 County and assize towns	63
3 Watering places (a) Inland	4
(b) On the coast	11
4 Seaports	
5 Manufacturing towns (a) Stockings	4
(b) Gloves	3
(c) Shoes	2
(d) Woollen goods	12
(e) Woollen and silk goods	1
(f) Silk goods	5
(g) Straw plait	2
(h) Cotton	12
6 Mining and hardware towns (a) Pottery	1
(b) Salt	3
(c) Copper and tin	7
(d) Coal	7
(e) Iron	7
(f) Hardware	2

Note: Twenty towns appear in more than one class.

shows. Now it has already been pointed out that such specialized centres were extremely vulnerable to change since once the highly particular *raisons d'être* for their new growth were removed for any reason, exhaustion of local point resources for example, or modification of technology, especially when it meant rationaliz-ation into large units, then the underpinning economy was destroyed. This argument can be extended into a universal principle, that towns come into being to perform specialized roles and that if the specialized role is removed then collapse follows, unless an effective general, regional-serving function has been developed. This was the crux of the translation of towns from medieval times into the modern; those which captured regional roles in administration and market become the metropoles, those that failed to effect such a capture faded and decayed to little more than villages, some even less than that where now an isolated castle marks where once a town was intended.

Once again, to provide an effective analysis of the structure of the urban size array, the period at which the process was complete needs to be identified. Here the answer is, perhaps, even more surprising in that the so-called de-industrialization of Britain in the 1970s marks the final phase of that process. Thus pit closures, and the 'destruction of communities', or, in central-place terms, their downgrading in status by a loss of function, are part of an ultimate process of adjustment. True that there is a greater or lesser politically based modification of that process, hence the plethora of regional and urban policies, but that does not disguise the fact that the later stages of the impact of industrialization upon the city system are being played out at the end of the twentieth century rather than at the end of the nineteenth century.

Empirical studies of the urban system

The next stage in this consideration of the changing structure of the urban system is the testing of the generalizations which have been made in attempting to answer the questions raised at the outset against the evidence of empirical studies. Unfortu-nately there are very few such studies to consider as against the much more frequent treatment of the rank–size array: there are many studies of the aggregated situation, few on local characteristics. The reasons are not difficult to assemble. The construction of an array from census populations is a relatively simple process, made all the more easy if C. M. Laws's tabulation of populations is used. The creation of local 'hierarchies', although the term is pejorative, based on counts of establishments of a variety of sorts is far more difficult and certainly more time-consuming. It has to be carried out by the use of trade directories, a process which is long, difficult and in the end gives results of questionable accuracy (Davies *et al.* 1968: Lewis, 1986; Shaw, 1982, 198). Moreover, as the urban geography of England and Wales in the nineteenth century became more widely researched, the main stimulus was the availability of the census enumerator books, a difficult but accessible source (see Chapter 2). This development of interest took place at a time when central-place studies were waning in popularity and social area analyses waxing. The urban geography of the nineteenth century became virtually domi-nated by the concern with social areas and the larger social geography derived from them. Central-place analysis has, therefore, been a minority interest reflected in the paucity of local, detailed studies.

The analyses of local urban systems in the nineteenth century are those by Carter (1956, 1965, 1969), Caroe (1968), Lewis (1970, 1975, 1986) and Barker (1978, 1980). The dates of publication are themselves indicative of the period in the late

1960s and the 1970s when some historical depth was being introduced into studies of the urban hierarchy. But of the published work, that of Caroe and Barker is largely concerned with methodological problems is establishing ranked arrays of towns from trade directory evidence, whilst that of Carter is based on a longer time scale within which the nineteenth century constitutes only a part. Again, all the studies apart from that of Davies are of rural rather than of the critical industrial areas. There are, therefore, clear limitations, but two of the areas can be used to give some illumination of local stability and change during the century. These are the analyses of Mid-Wales by Lewis and of the Rhondda Valleys by Davies, and since both are Welsh, a general outline of development within Wales can be prefaced.

Table 4.3 presents the rank order of the 10 largest settlements in Wales at each of the censuses during the nineteenth century. Straight census population figures have been taken and the considerable problems associated with the definition of areas, and the changing extent of urban boundaries, have been ignored. Thus, for example, at this stage, the Rhondda (the Ystradyfodwg Sanitary District until 1894 and Urban District until 1897), which was an amalgamation of a large array of small mining villages, is accepted and treated as one urban settlement, as it was in the census. But to call it a town would be nonsense; it was, as it was made in 1894, an Urban District.

In 1801 three constituents of the group making up the largest settlements can be identified. The first was Merthyr Tydfil which represented the new iron-working towns which had grown up in the latter half of the eighteenth century, the equivalent in Wales of the cotton towns of Lancashire. The second constituent was the two towns of Swansea and Holywell which represented a continuity and a revival of still earlier industrial, mainly metallurgical, centres in the south and the north. All the other towns were the traditional regional market centres which had functioned as such since late medieval times. Thus an older system of primarily regional serving market and administrative towns was overlain by a developing newer industrial set, the largest of which, Merthyr Tydfil, had no earlier urban functions of any sort. Of the five largest towns at the time of the Hearth Tax in 1670 all but one (Denbigh) appear in the 1801 list. There were minor changes in the early nineteenth century, some undoubtedly related to the fluctuating reliability of the census, but others to local and particular circumstances. For example, in 1851 Holyhead reached tenth place as a consequence of the building of the Menai Bridge in 1826. This also explains the appearance of Bangor in 1841 and 1851. Pembroke's entry in 1831 and its sporadic reappearance is closely related to the development and varying fortunes of Pembroke Dock. These changes, even in the top ten in Wales, perhaps indicate that the lower orders of the larger England and Wales array are being reviewed.

By the census of 1831 Cardiff and Newport, the two major ports, had entered the list and they gradually made their way up the table, Cardiff eventually becoming the largest Welsh town in 1881, following a standard line of progression, port – conurbation – metropolis. The growth of the coastal settlements was a direct consequence of the predominance of coal-mining in the interior valleys, especially after 1860. This is shown in the appearance of the Rhondda and Aberdare in 1871, together with other valleys' settlements which by that same year had displaced all the traditional regional centres; Carmarthen, which was the third-ranking Welsh town in 1670, fourth in 1801 and Caernarfon did not appear at all after 1861. Widening functional demands produced new settlements so that by 1891 Pontypridd, which became the service and shopping centre for the Rhondda, joined this set of largest towns. Barry, which reached ninth place in 1901, is an even more interesting example. Initially developed as an alternative coal port by a group of

Table 4.3 The populations of the 10 largest Welsh towns 1801–1901

1801	1811	1821	1831	1841
Merthyr Tydfil	Merthyr	Merthyr	Merthyr	Merthyr
7 705	11 104	17 404	22 083	34 977
Swansea	Swansea	Swansea	Swansea	Swansea
6 099	8 005	10 007	13 256	16 720
Holywell	Carmarthen	Carmarthen	Carmarthen	Holywell
5 567	7 275	8 906	9 955	10 834
Carmarthen	Holywell	Holywell	Holywell	Newport
5 548	6 394	8 309	8 969	10 492
Wrexham	Caernarfon	Caernarfon	Caernarfon	Cardiff
4 039	4 595	5 788	7 642	10 077
Haverfordwest	Haverfordwest	Haverfordwest	Cardiff	Carmarthen
3 964	4 327	5 271	6 187	9 526
Caernarfon	Wrexham	Wrexham	Newport	Caernarfon
3 626	4 234	4 795	6 000	9 191
Monmouth	Monmouth	Brecon	Haverfordwest	Bangor
3 345	3 503	4 321	5 734	7 232
Dolgellau	Brecon	Monmouth	Wrexham	Llanelli
2 949	3 303	4 164	5 484	6 846
Brecon	Dolgellau	Holyhead	Pembroke	Haverfordwest
2 705	3 064	4 071	5 383	6 601

entrepreneurs at odds with the control exercised by the Marquess of Bute at Cardiff, it quickly became a seaside resort drawing on the same industrial and urban hinterland it served as a port, and subsequently becoming little more than a suburb of metropolitan Cardiff; the 'mainly residential' town is thus found by the end of the century in south-east Wales. Of the ten largest towns in 1901 there were only two which had been in the 1801 list and one of those, Merthyr, was the total creation of industry and only there because of its very early development in the 1750s.

Before drawing conclusions from this brief review of the Welsh city system in the nineteenth century it is necessary to make some reservations. These relate to the fact that it is inadmissable to argue for a national status for Wales; it was essentially part of that amalgam called by the census 'England and Wales' and the borderland had no meaning in city-system terms. It follows that one is dealing with the lower parts of the rank-array and abstracting a regional example of a relatively minor order. When Cardiff became the largest city in 1881 its population was only 82,761 and reached 164,333 in 1901. Even so, it is worth noting the two critical conclusions. The first is that there was a total and fundamental revision during the industrial period. There was no continuity. The older and once significant regional centres were displaced by the new industrial towns based on the exploitation of localized resources of metal ores and coal and their processing and export. The second is that innovation played but little part in the sustaining of urban rank. Merthyr Tydfil was significant in the development of techniques in the production of iron and steel; above all it was the place where in 1804 the first steam locomotive ran. But that meant nothing when local iron ore was exhausted and the industry was faced with expensive import, including the uphill land transport from the coast of heavy, bulky raw materials. The industry collapsed to such an extent that in the 1930s an influential publication recommended that the town should be abandoned. In that

1851	1861	1871	1881	1891	1901
Merthyr 46 378	Merthyr 49 794	Merthyr 51 949	Cardiff 82 761	Cardiff 128 915	Cardiff 164 333
Swansea 21 533	Swansea 40 802	Swansea 51 702	Swansea 76 430	Swansea 90 423	Rhondda 113 735
Newport 19 323	Cardiff 32 954	Cardiff 39 576	Rhondda 55 632	Rhondda 88 351	Swansea 94 537
Cardiff 18 351	Newport 23 249	Aberdare 36 112	Merthyr 48 861	Merthyr 58 080	Merthyr 69 228
Holywell 11 301	Pembroke 15 071	Newport 27 069	Newport 38 469	Newport 54 707	Newport 57 270
Carmarthen 10 372	Llanelli 11 446	Rhondda 23 950	Aberdare 33 804	Aberdare 38 431	Aberdare 43 365
Caernarfon 9 883	Carmarthen 10 524	Llanelli 14 973	Llanelli 19 760	Llanelli 23 939	Pontypridd 32 316
Bangor 9 564	Tredegar 9 383	Abersychan 14 569	Tredegar 18 771	Pontypridd 19 971	Mountain Ash 31 093
Pembroke 8 977	Caernarfon 8 512	Pembroke 13 704	Ebbw Vale 14 700	Mountain Ash 17 495	Barry 27 030
Holyhead 8 863	Wrexham 7 562	Tredegar 12 389	Pembroke 14 156	Tredegar 17 484	Llanelli 25 617

sense it would have replicated the fate of medieval castle towns built to be regional centres but failing in the changed circumstances of later ages to survive.

This consideration of the city system can now be carried to a more local and detailed level. Two studies can be used. The first is that by C. R. Lewis of the Middle Borderland of England and Wales. Ranks were established not by population totals but from assemblages of functional elements derived from trade directories. Table 4.4 reproduces the critical changes which took place between 1828/31 and 1964/5. Apart from the one year 1868 the top four towns remained the same throughout the period, all of them on the English side of the border, stressing that there was no independent Welsh system. Indeed, Shrewsbury was larger than any Welsh town for the first three censuses, fell behind Merthyr only in 1841 and did not decline to the equivalent of seventh rank until the last quarter of the century. Even so, that larger comparison does indicate the significant element of change when the larger system, including the border towns of England, is reviewed. But in this area itself, where coal-based industry was of little revelance, there was the classic condition of stability at the top, some fluctuation in the middle orders and considerable change at the bottom of the selected array. As Lewis writes, 'these sudden shifts in status, both upwards and downwards, can be attributed to the vicissitudes of special activities at particular places' (Lewis, 1986, 117) but they were never of such magnitude to disturb the upper echelons. There, even, within the Welsh system, is the standard pattern of complete continuity in the larger centres from medieval times. It can also be noted that Barker's study of South West England between 1861 and 1911 reveals a similar situation (Barker, 1980).

The second study, that of the Rhondda Valleys by W. K. D. Davies, reveals very different processes at work. Figure 4.6 shows the number of commercial functions at the principal places in east Glamorgan in 1830 and 1849. Again, rankings are based on aggregates of functional elements derived from trade directories. This was

Table 4.4 Rank changes in 17 centres in the Middle Welsh borderland between 1828–31 and 1964–65 (after Lewis)

Rank	1828/31	1850	1868	1891/5	1964/5
1	Shrewsbury	Shrewsbury	Shrewsbury	Shrewsbury	Shrewsbury
2	Hereford	Hereford	Hereford	Hereford	Hereford
3	Leominster	Ludlow	Ludlow	Ludlow	Ludlow
4	Ludlow	Leominster	Newtown	Leominster	Leominster
5	Newtown	Newtown	Leominster	Newtown	Welshpool
6	Kington	Welshpool	Welshpool	Welshpool	Newtown
7	Welshpool	Kington	Kington	Llanidloes	Knighton
8	Hay	Hay	Hay	Kington	Church Stretton
9	Llanidloes	Llanidloes	Knighton	Builth	Hay
10	Presteigne	Knighton	Builth	Knighton	Kington
11	Bishop's Castle	Builth	Bishop's Castle	Bishop's Castle	Builth
12	Knighton	Bishop's Castle	Llanidloes	Hay	Llanidloes
13	Montgomery	Presteigne	Rhayader	Presteigne	Rhayader
14	Builth	Rhayader	Llanfyllin	Llanfyllin	Bishop's Castle
15	Llanfyllin	Montgomery	Presteigne	Montgomery	Presteigne
16	Rhayader	Llanfyllin	Montgomery	Rhayader	Llanfyllin
17	Church Stretton	Church Stretton	Church Stretton	Church Stretton	Montgomery

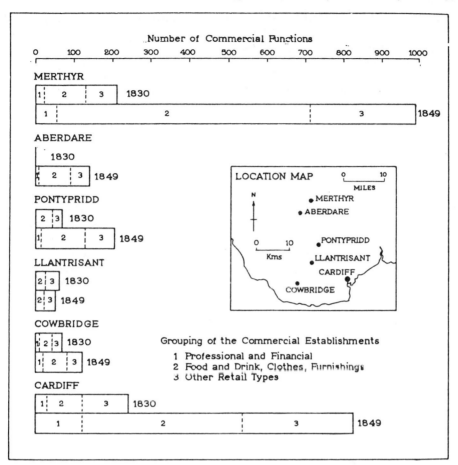

Figure 4.6 Numbers of commercial functions at principal places in east Glamorgan, 1830 and 1849.
(after Davies, 1970, 198, Figure 9.3)

at a period when the iron-working towns of the northern outcrop of the coalfield, such as Merthyr and Aberdare, had been established for nearly a century, but before the great boom period of coal mining which was to characterize South Wales after 1860. Even of this early period, immediately after 1860, Davies writes, 'Pontypridd clearly outpaced the older market centres, and Llantrisant actually declined in the period, reflecting loss of part of its hinterland to Pontypridd' (Davies, 1970, 198). Thus the medieval towns of Llantrisant and Cowbridge were losing out before the advance of new industrially based settlements. These newer towns were becoming divided, generally, if never absolutely, into those where point production was predominant and those at strategic valley-confluence locations where service roles were significant. Merthyr, at the head of the Taff valley, was just beginning to be challenged by the down-valley confluence settlement of Pontypridd. By 1830, on the basis of a functional measure, Cardiff was already nearly parallel with Merthyr Tydfil and by 1849 the greater professional and financial component at Cardiff was indicative of the future and of the advantage of the 'ultimate' confluence and gathering point of the area.

Figure 4.7 charts the changes in functional significance of the larger settlements in the Rhondda Valleyes after 1868, that is during the period of the vast increase in coal mining which raised the population from 3,025 in 1861 to 113,735 in 1901. The new entrants are the larger mining settlements which, of necessity, provided local and immediate services. By 1901 the population of Pontypridd was 32,316 and it was ranked seventh in the array of Welsh towns. However, as Figure 4.7 shows, the initial impact of growth was to lower the relative dominance of the town as the smaller mining towns created their own service arrays, but it was to rebound in the twentieth century. What this discussion reveals is the standard process of towns, or perhaps settlements is the better term since many were but 'urban villages', being created for highly specialized purposes, in this case coal mining. But the accumulated populations have to be served. At first, this, at its crudest level, was done by the truck system, by the company store, especially in remote areas. Eventually, however, an integrated and relatively stable system emerged from competition and in that, locational advantage seemed to have been at least as important as inherited status and past control. Clearly by the end of the century, although the major lineaments of the industrial rank array were apparent there were still changes in process, and indeed it can be argued, as it was earlier in this chapter, that they have continued well on to the twentieth century and are still a central concern of regional planning. There is no question, however, that the major disruption took place between 1861 and 1881. Figure 4.8 shows rank correlation

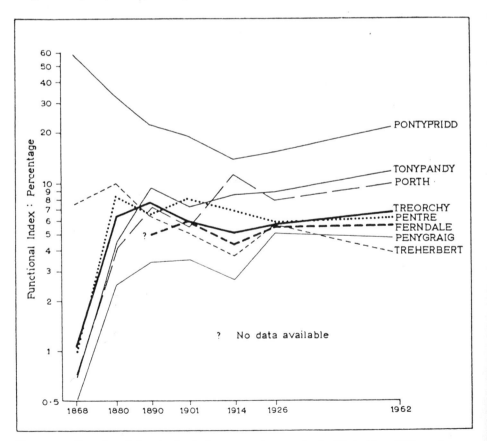

Figure 4.7 Functional changes in the largest settlements in the Rhondda, 1868–1962. (after Davies, 1970, 206, Figure 9.6)

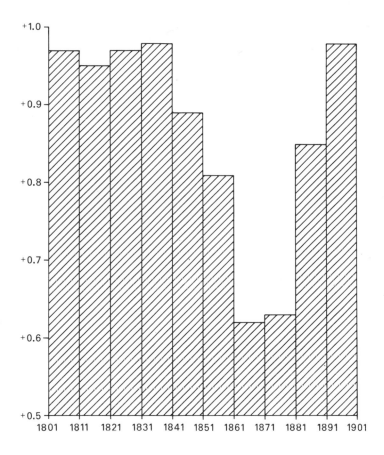

Figure 4.8 Spearman's rank correlation coefficients for rankings of Welsh towns for each decade of the nineteenth century.

coefficients for the ranking of Welsh towns for each decade of the nineteenth century and the low values for those two decades indicate both the date and the extent of the shake-up.

This review of Welsh towns can now be turned back to consider the complete arrays in 1801 and 1901 as set out in Figure 4.9. The first and most obvious point is the increase in the number of towns. Using census definitions again which refer to administrative areas and not physically distinct settlements, that is the Rhondda becomes one settlement, the urban total increased from some 61 in 1801 to 111 in 1901. If a more realistic definition were used the increase would be revealed as much greater. Along with this went another immediately apparent feature, a marked growth in the size of towns. In 1801 Merthyr was the largest at 7,705; by 1901 Cardiff had reached 164,333. When, however, the rank–size arrays on a log-normal basis for the two dates are compared, no great differences in the shapes of the resultant curves appear in spite of a period of apparently revolutionary change. It seems that some form of allometric process was in operation by which growth in the number of settlements was compensated by increases in the sizes of the largest towns. But given the arbitrary nature of the population figures, it is hardly worth carrying the discussion beyond that general statement.

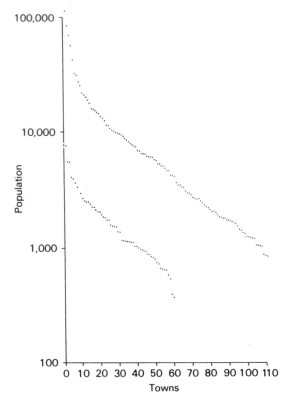

Figure 4.9 Rank-size distributions for Wales at the beginning and end of the nineteenth century.

Conclusion

This conclusion returns the discussion to the question posed at the beginning of this chapter which has only been partly considered. What was the nature of the city system itself and can any changes be observed during the century? In general terms it is possible to contend that city systems do move through identifiable stages (Carter, 1983). The earliest and most primitive is marked by minimal transportation facilities and an undeveloped economy. Interaction is severely limited and the separation of what centres there are means that no systematic relations between them exist. Development, usually political in character, results in one centre capturing control and emerging as a primate city when it dominates a set of much smaller relatively undifferentiated centres. Further economic growth, especially within the regions removed from the primate city, converts the system into one more nearly like the hierarchical ordering envisaged by Christaller with the provincial capitals forming the second-order cities. Finally, industrialization, and with it a great widening in the range of functional specialization as the bases of towns, blurs the underlying hierarchy and converts the system into an approximation to the rank–size rule.

Figure 4.10 sets out the rank–size array of towns in England and Wales for each census year from 1801 to 1911. It is difficult to identify in these graphs any progressive pattern of change which would reflect those outlined above. Certainly

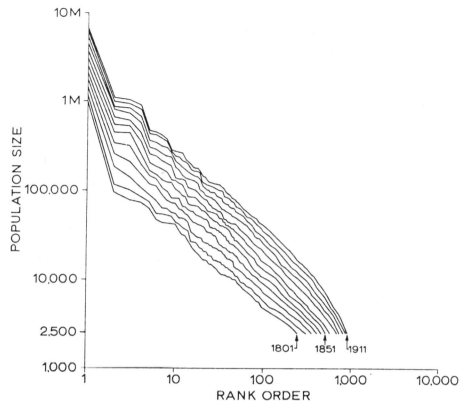

Figure 4.10 Size arrays of cities in England and Wales, 1801–1911.
(after Robson 1973, 30, Figure 2.3)

London at the start of the period was still the primate city which it seems to have been before 1750. Table 4.5 compares the population of London with the next largest city in England and Wales between 1600 and 1901. The multiplier in 1600 at 17 is an unequivocal reflection of primacy.

During the period after 1600, however, the multiplier progressively declined to 6 in 1851 and remained thereabouts, depending on the definition of areas, for the rest of the century. That still indicated primacy since a multiplier of 2 would have been needed to indicate that the rank–size rule was applicable. It seems that the late date of the first census in relation to the onset of industrialization meant that any hierarchical ordering had been blurred by 1801. It is difficult to perceive any national principle other than the primacy of London, but that was certainly not accompanied by an extensive set of equally ranked towns. As already suggested,

Table 4.5 The populations of London and the next largest city from 1600 to 1851 (after Daunton)

Date	London	Next largest city	Multiplier
1600	250,000	15,000 (Norwich)	17
1750	655,000	50,000 (Bristol)	13
1801	960,000	84,000 (Manchester)	11
1851	2,400,000	376,000 (Liverpool)	6

throughout the century local hierarchies were merged under London to produce a system which only very roughly approximated to a rank–size rule. Like most attempts to discern order in the totality of nineteenth-century England and Wales an unsatisfactory conclusion is reached. It might well be that the perception of size and significance by entrepreneurs in retailing and the providers of services was to create the endowments of cities, the 'trait complexes' which identified hierarchical orders and the clear establishment of such systems was a feature of the twentieth century, rather than the nineteenth.

To a degree, if only to a degree, the nature of the evolution of the city system in the nineteenth century reflects the conflict of views as to whether industrialization suppressed or accentuated regional differences in England and Wales (Langton, 1984, 1988; Gregory, 1988a, 1988b). In terms of the city system Langton's views have significance. What industrialization initially created was a series of independently organized regional systems which, during the century, were very gradually synthesized by the development of the capitalist system, in all its aspects, into a single interacting system under the metropolitan dominance of London. But that state was only reached towards the end of the century. Black, after his examination of financial capital in early industrial England, concludes, 'In this context a regionally fragmented series of export-based economies bound together by metropolitan–dominated flows of finance capital and commercial information is perfectly plausible' (Black, 1989, 381). That probably represents fairly the state of the system, if such it can therefore be called, during much of the nineteenth century and

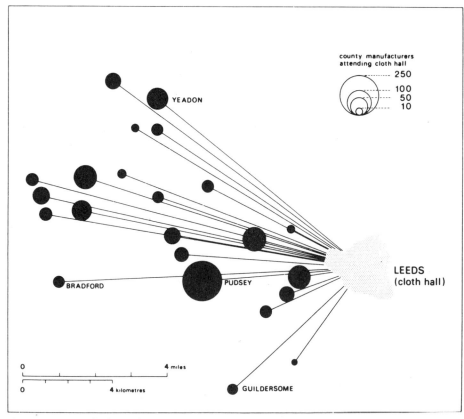

Figure 4.11 A Cloth Marketing Systems, Leeds.

until, under the twin effects of increased mobility and effective inter-regional competition, an integrated system emerged. The implication of course, is that to construct rank–size graphs over much of the century and arbitrarily unify a series of independent systems is a futile exercise. Langton, in this respect, is right. What is needed is the establishment of the nature of the regional systems and the process of their integration as nation-wide retail and commercial operations came into being.

It is extremely difficult to trace the establishment and development of these separate regional systems. They probably reflected regional markets in the staple commodities and in that context Gregory's mapping of the markets in the cloth industry of Yorkshire provides the best insights (Gregory, 1982). Those who made the journey to the cloth market were creating affiliations which would be translated

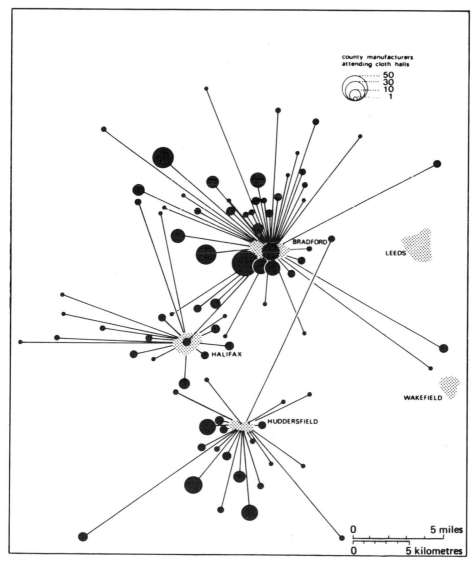

Figure 4.11 B Bradford, Halifax and Huddersfield.
(after Gregory, 1982, 117, Figure 3.12 and 118, Figure 3.13)

into the more conventional and universal demands made on central places as the century progressed and new administrative and shopping needs emerged. And there was competition. As against Leeds 'other major markets existed: Bradford built a Piece Hall in 1773; Halifax had a Cloth Hall very early in the eighteenth century and a new Piece Hall by 1779; Huddersfield built a Manufacturers' Hall in 1766; and Wakefield replaced its old Cloth Hall with a Tammy Hall in 1776. But the two Leeds Halls dominated the woollen trade' (Gregory, 1982, 115). The maps in Figure 4.11 show the overlapping cloth marketing systems in 1822. It was from these bases that the system of linkages was built up and competition between markets resolved into a regional central place system under the regional leader, in this case Leeds, just as Cardiff became dominant in South Wales. Integration into a national system probably came first in the financial world and was forwarded by the organization of local and national government. The transformation of retailing during the century also promoted unification into a single system capable of being interpreted on the principles set out by Christaller. But it was near to the First World War before such an interpretation can be accepted.

The conclusion to this chapter must be that there was a revolutionary overturning of the city system inherited from the Middle Ages and the mercantile impacts of the pre-industrial period. True, there were backward links but what is significant is the massive changes which were engendered. Even London with a population of 7.2 million in 1911 was a vastly different entity from that with only 250,000 in 1861. Completely new forces had come into play and with them the internal structures of cities were transformed. It is to those internal transformations that the remainder of the book is directed.

5

Intra-urban change: the processes of restructuring

The remainder of the book deals with the transformation of the internal structure of the towns of England and Wales which took place during the nineteenth century. That constitutes a formidable task given the complexity of the processes operative and the great variations in the size and functional character of towns. Moreover, as the Introduction set out, there is the ever present problem of venturing into every aspect of city life for all are interlocked and germane. But some progress can be made by arguing that urban structure was the outcome of the interplay of three sorts of process – economic, social and political – operating within environmental conditions and constraints. This is set out in diagrammatic form in Figure 5.1. The sub-headings indicate the most significant aspects within each process type. They are not exhaustive, neither are they independent. Thus for example, class structure which was essentially part of the social system, was a product of occupational diversity which was, in turn, derived from primarily economic bases. Here, once again, the issue of reductionism arises. In this book these processes are accepted as operative and not traced back *in extenso* to their origins as part of the larger-scale development of the capitalist system. Even so, it is necessary to consider the nature of the class system in interpreting residential segregation. In the same way it is essential to examine the movement for public health reform which as the precursor of the town planning movement had such an impact upon urban character.

It would be a feasible approach to take each of the sub-headings in Figure 5.1 as the basis of the investigation of internal town structure. But that would inevitably result in a good deal of repetition because of the interrelations which have been noted. Moreover, if the geographical discipline of spatial patterning is to be stressed then the emphasis must be on that, rather than on the aspatial processes. Such a principle would predicate a structure for the book based upon the various distinctive regions which made up the towns. That is indeed the basis which is adopted. Thus the major towns of the nineteenth century were industrial towns and it is logical to begin with industry as the significant new user of urban land.

The phrase used above – significant new user – stresses the critical change which industrial capitalism brought about. Land in cities became a commodity to be bought and sold for profit, not held for reasons of prestige and status. It is that which lies at the foundation of the analysis of the differential bids for land which were made by competitors. The chapters which follow, therefore, take each of the major bidders in turn and they examine the nature of the claim on urban land and the locational characteristics which followed. Again, it is a complex process for users, conveniently aggregated into simple categories, such as industry or retailing,

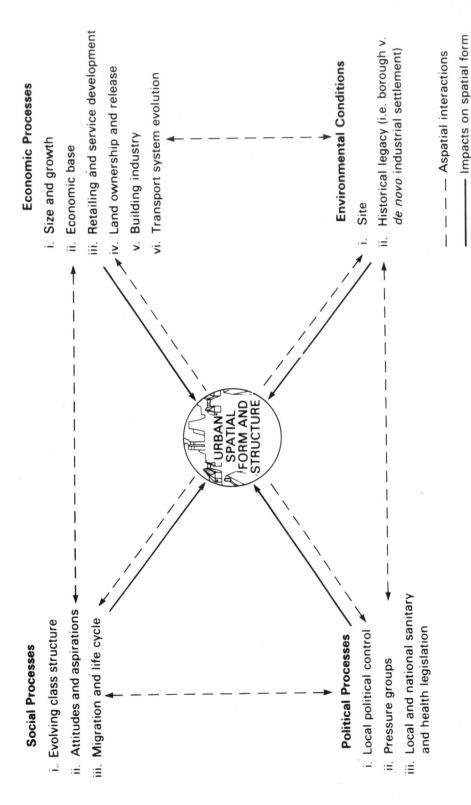

Social Processes

i. Evolving class structure
ii. Attitudes and aspirations
iii. Migration and life cycle

Economic Processes

i. Size and growth
ii. Economic base
iii. Retailing and service development
iv. Land ownership and release
v. Building industry
vi. Transport system evolution

Political Processes

i. Local political control
ii. Pressure groups
iii. Local and national sanitary and health legislation

Environmental Conditions

i. Site
ii. Historical legacy (i.e. borough v. *de novo* industrial settlement)

- - - Aspatial interactions

——— Impacts on spatial form

URBAN SPATIAL FORM AND STRUCTURE

Figure 5.1 Urban spatial form and structure: summary of processes. (after Carter and Lewis 1983, Figure 1.5)

were far from homogeneous in their demands so that within each category different solutions were sought or accepted. No more so was this true than in residential use where the elaborating class structure resulted in demands which were satisfied, or were not satisfied whichever the case might have been, in widely varying ways and locales.

In the conventional bid-rent theory interpretation urban structure is seen as the outcome of the resolution of competing demands for limited urban space. Who bids highest wins. Those who can make no effective bid are relegated to the least desirable locations; in absolute residential terms they become homeless. But bids are not the product solely of the evaluation of situation for the amount of land required for any operation is a factor. This free market bidding system interpretation is appropriate to the nineteenth century, indeed it can be regarded as the apotheosis of industrial capitalism. But there never was a free market. Suppliers could operate policies which limited the free availability of land and were able to put limiting conditions at the time of sale. In addition throughout the century there were political constraints, as well as public pressures, upon both users and suppliers.

The supply side breaks down, as it does today, into the threefold impact of land owners, developers and builders. Certainly the nature of the three was very different from the present, but the willingness to release land through sale, the organization of development and the actual process of erecting buildings all had a significant impact upon the townscapes which resulted.

As the century progressed yet another influence became more apparent. In Figure 5.1 it is simply labelled 'political' but it embraces all those constraints which arose not only from government legislation, often as it was locally applied, but also those which were the result of public pressure. Philanthropy, and the very name has a Victorian ring to it, was important not only through its own developments but in the impact which it had on political action and legislation. All the private efforts to offset the worst extremes of the operation of the free market come rather unhappily under a heading of political processes but many, if not most, were ultimately aimed at encouraging political action via legislation and are legitimately considered as limitations upon the operation of demand and supply. But the central theme is the development of effective local government, culminating in the 1880s with the creation of the system which was to last until 1974, and its operation of an increasing system of constraints, mainly directed towards public health and sanitation. During the whole century, however, the influence of urban managers and pressure groups was a cogent factor within the processes of urban growth.

The remaining chapters of the book are based on the foregoing discussion. The demanders of urban land are taken first since they are the proactive forces. Industry and transport, commerce and government, institutions and public utilities are taken in turn, concluding with residence, which became the largest consumer of land. The suppliers of urban land and the builders, who used it, are considered next, and then the controls and constraints which by the beginning of the next century were to lead to the first Town Planning Act of 1909.

6

Demands for urban space: industry and the railways

It is paradoxical that one of the least well treated aspects of urban land-use in the industrial city of the nineteenth century should be industry itself. Certainly few geographical studies review intra-city industrial use of urban land in the nineteenth century at any length or in any depth. A recent commentator (Shaw, 1989, 55) has remarked on 'the fragmented nature' of research on the economies of industrializing cities, and of the 'relatively low level of investigation and few studies on industrialization in nineteenth-century cities'. This is certainly so, for most of the well known studies (Gregory, 1982; Mounfield, 1967) have been concerned with distribution on a regional or national scale, rather than with specifically intra-urban patterns. Even the traditional descriptive models of urban form are mainly directed towards the differentiation of residential districts; the more sophisticated analytical models are to an even greater degree related to housing. All this is probably a response to the availability of data and to intellectual fashion, the two not unrelated. But if the epithet 'industrial city' means anything it is surely that the driving force which made it different from other cities was, quite obviously, industry, and that sets up the assumption that the land devoted to industry performed the initial organizing function within the mosaic of land uses.

Against the above background the actual evolution of industrial land-use during the century in the most general of contexts is not difficult to set out. The pre-industrial situation was one where manufacture was for a local market, small-scale in nature and took the form of craft workshops which, although in larger cities and in specialized trades became spatially associated and took on specific locations, were generally distributed within an undifferentiated town centre. Industrialization was based on the factory, the crucial element in the social and economic transformation, but also in the urban transformation. Locational controls, related to power sources on the one hand and the distributive network on the other, became much more demanding and much larger areas of urban land were needed. Both forward and backward linkages, which became more complex during the century, exerted pressure to connect the single free-standing mill or factory into an industrial complex occupying even more extensive tracts within the town. Model towns or industrial villages were built about industrial installations. Further changes were to associate lighter industries on industrial estates, but these were to emerge only after the century's end, and more particularly during the inter-war years. But the workshop basis for many industrial activities never disappeared and remained distinctive well into the twentieth century. Indeed, one of the main reasons for its demise was the destruction by bombing during the Second World

War and the comprehensive redevelopment of the inner city which followed it. As Peter Hall (1962, 21) writes, 'much of the industrial production of Britain [at the end of the nineteenth century] still came either from small factories and workshops, or from larger units that specialized in one-off bespoke craft production as in the Tyne and Clyde shipyards ... This was the characteristic mode of industrial production up to the 1880s, not merely in London but elsewhere. The great metropolitan cities were dominated by small-scale artisan production, mostly catering for the demands of the metropolitan community and hidden away in small-scale workshops.'

Craft industry in the nineteenth-century city

There have been two classic surveys of the distribution of craft-based manufacturing in British cities during the nineteenth century, although in both cases the historical evidence was a prelude to an analysis of contemporary patterns. One of these was Peter Hall's study of *The Industries of London since 1861* (Hall, 1962), and the other was Michael Wise's paper 'On the evolution of the jewellery and gun quarters of Birmingham' (Wise, 1949).

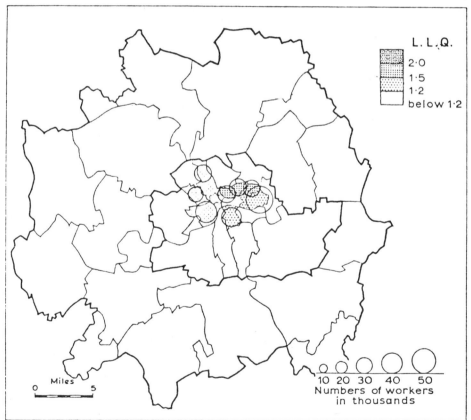

Figure 6.1 Manufacturing industry in Greater London 1861. Boundaries are 1861 Registration Districts.
LLQ is London Location Quotient.
(after Hall, 1962, 29, Figure 1)

The general distribution of manufacturing industry in Greater London in 1861 is shown in Figure 6.1. It made up a sector which ran from the City to Westminster and an extension south across the Thames to Southwark, towards the northern and eastern boundaries. Two blank sectors, the City and Holborn and Islington broke into it, thus creating a horseshoe shape constituted by two independent centres, one in the West End, the other in the East End. The pattern is replicated in the specific detail of the clothing industry which Hall examined, along with furniture making and printing. One aspect, tailoring, can be considered here, shown in the distributions set out in Figures 6.2 and 6.3. Westminster and Stepney accounted for nearly 40 per cent of the tailoring workers of Greater London in 1861. The West End area was bounded by Bond Street, Oxford Street, Regent Street and Piccadilly and at its heart, epitomizing concentration and specialization, were the streets such as Savile Row, which were then and now universally associated with tailoring. The East End area in Stepney was mainly focused on the western margins in Whitechapel. The reasons for these locations bear little relation to those traditionally advanced for industrial sites. Raw material sources and the availability of fuel and power were of little moment, as indeed they were to the craft industries in general. The critical element was the market, and whereas in the small town the demand by those highest in the social scale produced a number of tailors scattered in the central area, in London the vastly greater numbers, and the vastly greater wealth of those numbers, produced a high concentration in the West End which had long been the residential area for aristocrat and plutocrat. Also crucial in this market relationship is fashion. The close association of customer and producer enabled the demands of fashion to be met instantaneously. The East End concentration was somewhat different and depended largely on the development of ready-made clothing for a

Figure 6.2 Distribution of tailoring in London's West End in 1861. Mapping is not continued north of Marylebone Road or west of Edgeware Road.
(after Hall, 1962, 41, Figure 5)

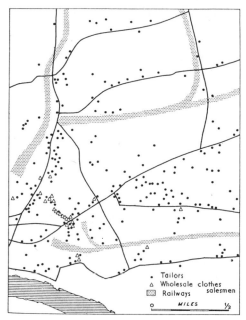

Figure 6.3 Distribution of tailoring in Stepney and Bethnal Green in 1861. (after Hall 1962, 43, Figure 7)

mass market after 1850, it being made viable as the consequence of innovations in cutting and sewing machines. Sub-contracting, or sweating, characterized an industry which remained fundamentally small-scale. 'In clothing, then, the Industrial Revolution failed to engender a factory system; on the contrary, it led in the second half of the nineteenth century to a more and more vertically disintegrated system of production in which a sweater might sub-contract the specialized jobs of pleating, button-holing, button-covering and embroidering to firms in his area' (Hall, 1962, 55). Bythell, in his book *The sweated trades* which provides an extended account of these, quotes a London tailor's machinist, 'it is a remarkable thing how these people are unacquainted with the whole of the trade; the person who presses a pair of trousers could not press off a coat, the person who presses a waistcoat could not press off a coat, but the one who presses a coat could not press off a waistcoat' (Bythell, 1978, 67). But the average tailoring firm in England and Wales at the 1861 Census was only 3.2 workers.

The other locational factor, especially relevant to the development of ready-made clothing, was labour supply. There were two distinct elements within it, women and Jewish immigrants, both of which could be exploited by minimum payments and by the progressive deskilling of operations. Although the scale of the clothing industry increased enormously in the nineteenth century its basic organization remained the same. Based on small workshops, it merged into the general characteristics of the contemporary inner city rather than creating distinctive industrial complexes. Even the mass-produced clothing of an entrepreneur such as Montague Burton at Leeds was to develop only after 1900.

The metal industries of Birmingham present a somewhat different process of evolution (Wise, 1949). Although based, at least initially, and during most of the nineteenth century, on small workshops, they were high specializations already involved in international markets. Unlike tailoring, they were not commmon to most cities, but were uniquely the products of Birmingham. The metal trades arose out of the decay of earlier products such as edge tools, leather and textiles, and as a response to the development of coal mining and iron working in South Stafford-shire. Birmingham's response was to concentrate on products requiring high skills and limited raw materials. In this respect, characteristic was gun manufacture and assembly from semi-finished parts produced in the new manufacturing towns of Wednesbury and Willenshall. Wise notes that gunsmiths were already men of esteem in Birmingham in 1700 but that locational concentration came about after 1740 with the construction of the Birmingham canal, with the development of estates which lay to the north-west with easy access to the Black Country. To the south, restrictive covenants excluded industry (Chalklin, 1974; Cannadine, 1980), whilst to the south-east the low lying alluvial land was liable to flooding. Gun-smiths had spilled over northwards onto the area about St Mary's church (Figure 6.4) by the mid eighteenth century and were firmly established there by 1825. Distributions in 1845 and 1865 (Wise, 1949, 70) indicate an increase in extent but no change in location. The reaction to large-scale military demand for guns was to create the Birmingham Small Arms (BSA) factory in 1862 in the middle industrial ring, but competition from larger-scale production, as at Enfield, was to lead to even greater specialization on skill and the manufacture of expensive sporting guns.

A similar pattern of growth characterized the jewellery industry, but with less stability of location. It derived from the making of buckles and metal toys which were the basis for the development of higher-quality jewellery led by the Soho factory of Matthew Boulton. An assay office for silver had been established by

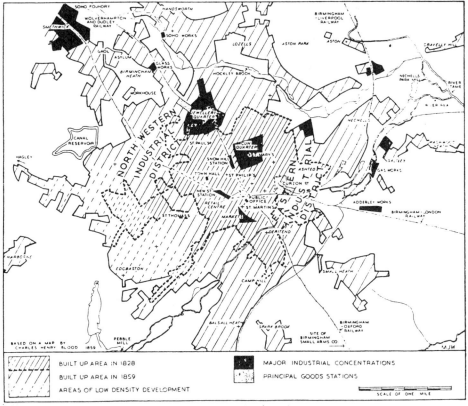

Figure 6.4 The distribution of industry in Birmingham in 1859.
(after Wise and Thorpe, 1950, 221, Figure 43)

1775. It was between 1800 and 1830 that a clear concentrated location can be discerned, again as part of the movement north-west. The Colmore estate had become available for builders under a private act of 1746 (Chalklin, 1974, 82) and it led to expansion so that jewellers associated immediately to the north-west of St Paul's church, and especially on the Newhall estate. By 1845 the trade employed some 3,700, whilst there were about 5,300 toymakers, many of whom probably made jewellery (Wise, 1949, 69). After 1865 and the discovery of gold in California, that metal became part of the trade, and another estate, the Vyse estate, had been opened up and the conversion of houses into manufactories followed. 'Many of the premises on the Vyse estate were larger than the average north Birmingham house-workshop of the period, and an actual migration of firms can be traced' (Wise, 1949, 70) into the new area. The centre of location became firmly fixed in Vyse Street and Northampton Street and remained there for the rest of the century.

To gun making and jewellery there should be added another Birmingham specialization, the brass trade, which employed some 8,334 persons in 1861. Its rapid expansion was in response to demand arising both from engineering and domestic purposes. Again, before 1825 'the trade was carried on entirely in "small workshops, low roofed and imperfectly lighted . . . for the most part situated in back courts". But after that date new larger factories became general and brass working characterized the industrial areas of the north and north-west of the town and also the north-eastern and eastern industrial districts' (Wise and Thorpe, 1950, 216).

The establishment by BSA of their factory at Small Heath in 1861–62 was indicative of the change both in structure, from workshop to factory, and in location, from inner city to a 'middle belt' of industry, which was taking place in the latter part of the century and creating a series of nodes, such as Bordesley, Saltley and Nechells. Part of this development, although shifting even farther away from the centre, was the transfer by the Cadbury brothers of their cocoa and chocolate factory from the city centre to Bournville.

In terms of city structure it will be apparent that at Birmingham there is an immediate approximation to the simplest of all urban models. An inner city area demonstrated by 1925 the characteristics of Burgess's zone in transition, for by then modern forces were beginning to act upon it although major redevelopment only occurred after the Second World War. The middle belt, broken as it was by such elements as the Calthorpe Estate in Edgbaston, was very much the zone of working men's housing and factory industry. It is, perhaps, this conjunction which has given rise to the tendency to ignore specific industrial elements in the discussions of industry in industrial towns.

Factory industry in towns

The situations examined so far have been derived from the intensification and greater specialization of the types of crafts which were present in the pre-industrial town. If the types of heavy industry which epitomized the nineteenth century are considered, then somewhat different conclusions follow. Three Welsh towns can be briefly used in exemplification. Neath, in the county of Glamorgan, was a moderately sized town which grew from 2,512 in 1801 to 13,720 in 1901. But these totals exclude its southern extension in Briton Ferry which was returned as a separate Urban District between 1871 and 1921. The combined population in 1901 was 20,693. The location of industry was closely related to the main lines of transport (Figure 6.5) where the River Neath, the Neath Canal (1791) and the South Wales Railway (1850), along with the main road from Cardiff to Swansea

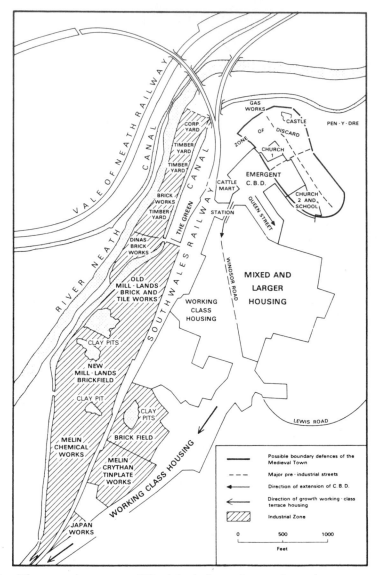

Figure 6.5 The general structure of Neath in the later nineteenth century.

and the west which diverted north to cross the river, made up an axis along which the major industrial components were located. These were tinplate and chemical works, together with brick and tile works which had developed along the river from an initial location near to the town quay. This elongated zone extended southwards to the river mouth at Briton Ferry and included railway repair shops and, at the docks, the Briton Ferry Steel Works. It was this zone which was the organizing force within the growing town. A series of working-class residential areas, virtually a series of urban villages, grew in association with it so that the bulk of the growth was south towards the navigable part of the river (Figure 6.5).

These developments had a significant impact upon the town. Population growth resulted in an increase of retail, trading and professional functions and, as that took place, so the locations of those activities shifted south and away from the historic core. High Street, the main axis of the medieval and post-medieval town and where the central market hall was located, became no more than a decaying part of the shopping area, a true zone of discard into which the poorest and least skilled section of the population filtered. Even by mid-century what would now be called a characteristic inner city area had been created. As a corollary, those highest in the social scale began to move away to higher ground to the east, although always under the constraint of limited mobility. By 1881 it had pushed still farther outwards when already a small number of detached villas in their own grounds had been built along Lewis Road (Figure 6.5) and on to Westernmoor.

As a comparison with Neath, its neighbour to the west, Swansea, can be briefly analysed in the same way (Figure 6.6). It was in origin, like Neath, a medieval borough which by the eighteenth century had developed two special functions. Its seaside location had generated a resort role, whilst in the early eighteenth century, initially at Landore in 1717, metallurgical industries had developed in the Tawe Valley to the north of the town. By the mid nineteenth century the industrial base

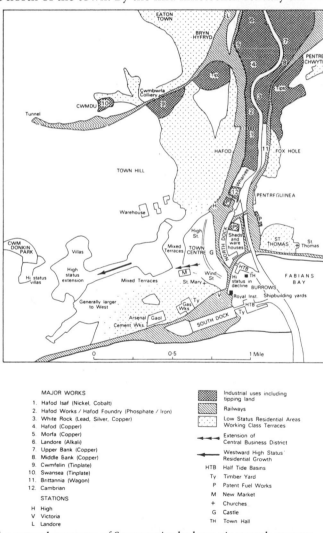

MAJOR WORKS

1. Hafod Isaf (Nickel, Cobalt)
2. Hafod Works / Hafod Foundry (Phosphate / Iron)
3. White Rock (Lead, Silver, Copper)
4. Hafod (Copper)
5. Morfa (Copper)
6. Landore (Alkali)
7. Upper Bank (Copper)
8. Middle Bank (Copper)
9. Cwmfelin (Tinplate)
10. Swansea (Tinplate)
11. Brittannia (Wagon)
12. Cambrian

STATIONS

H High
V Victoria
L Landore

Industrial uses including tipping land	
Railways	
Low Status Residential Areas Working Class Terraces	

Extension of Central Business District

Westward High Status Residential Growth

HTB   Half Tide Basins
Ty   Timber Yard
P   Patent Fuel Works
M   New Market
+   Churches
G   Castle
TH   Town Hall

Figure 6.6 The general structure of Swansea in the later nineteenth century.

was dominant and the population increased from 6,099 in 1801 to 94,537 in 1901. Industry progressively grew along the flatter valley land of the Tawe to the north, dominated especially by a distinctive group of copper works – Morfa, White Rock, Middle Bank and Hafod (Figure 6.6) together with a wide range of other metallurgical industries and the manufacture of patent fuels and chemicals. As at Neath the result of this industrial growth was to create a distinctive extent of working-class housing in close proximity, so that a sheath of railway and industrial land followed the river, bordered and inter-penetrated by the lowest quality housing. The old city centre became a zone of discard as at Neath, but the main growth was westwards and away from the riverine industrial zone. High quality housing developed in sympathy along the lower slopes of Townhill generating a west–east contrast which has remained to the present.

Some general conclusions can be drawn from the consideration of these two towns.

1   In both towns industrial growth created a distinctive elongated zone, closely associated with the river front and the main communication lines. Because of contrasts of site and situation the zone lay to the south of the town in Neath, to the north in Swansea.
2   Because walking was the only means of getting to work, a series of industrial villages housing the working-class population was intricately linked with the industrial zone in both towns.
3   In both towns, one-time prestige areas of central residence experienced decline as industry became dominant and the central area expanded. High status residence was displaced, the actual buildings being converted to other uses.
4   The area of high status residence, as it was displaced, extended away from the centre and especially from industry which was thus crucially responsible for the creation of socially contrasted residential areas.
5   The shift of the central business area created in Neath a clear zone of discard; in Swansea Wind Street, once the fashionable shopping street, lost its pre-eminence. Again the direction of industrial growth impacted upon the growth and character of the central area.
6   In both towns the poorest areas became located on the margins of the central business area as former uses moved away either into newer retail areas or to the developing prestigious residential areas.

It is clearly difficult to treat the industrial areas of Neath and Swansea without also considering characteristic land-uses which form part of other chapters – retail and residence are the best examples. But this is simple testimony to the significance of industry in organizing the basic structure of nineteenth-century towns. This is even more apparent in the third example, Merthyr Tydfil, which was the creation of industry, there being no previous settlement other than a small village. It was the largest Welsh town both in 1801 at the first census at 7,705, and at mid-century with a population of 46,378 in 1851. The forces generating urban growth were the iron-works complexes. Although it is a simplification, there were four such complexes which brought distinctive industrial regions into being at Dowlais, Cyfarthfa, Penydarren and Pentrebach (Figure 6.7). These regions were made up not only of the works themselves but also extensive areas bought up by the iron masters and used for the dumping of waste from furnace and mine. Closely associated with the iron works, and often hemmed in or overlooked by spoil heaps, were working-class residential areas: they, again, were pre-eminently urban villages.

The demand derived from this rapidly increasing population resulted in the

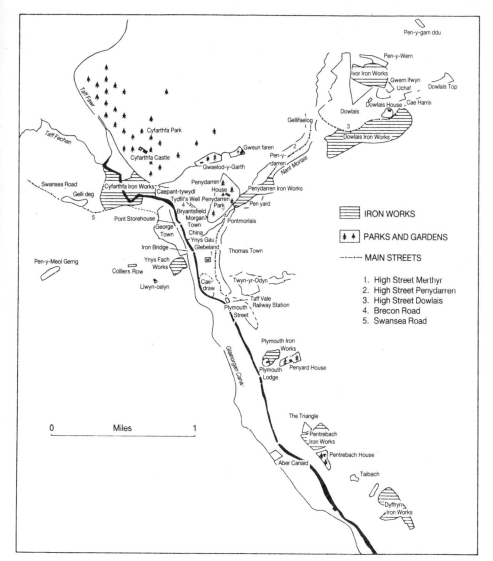

Figure 6.7 Merthyr Tydfil in 1851.

creation of a distinctive shopping and business centre. It was probably initially associated with the old village but soon grew rapidly away to the north along what was made into High Street (Figure 6.7). It was pulled in that direction by the centre of gravity of population which lay to the north where the main lines of movement along the Taff valley converged with that along the Morlais valley from Penydarren and Dowlais. It was fixed by the building of a market hall after the Merthyr Market Act of 1835.

Merthyr is a classic multiple-nuclei town with its four iron-working settlements (Dowlais can almost be treated as a separate sub-town) focused on the fifth nucleus, the central area. That central area was in itself greatly complex and still reflects the basic localization of industry. The Ordnance Survey map of 1878 for example still shows breweries, tanneries, corn mills, smithies, a whole variety of smaller-scale factory and workshop industries within the central area. Pigot and Company's Directory for the earlier date of 1830 showed six tailors, five dressmakers and two hatters, even so very modest numbers compared with the 12, 13 and three at the more urbane Swansea. But the whole structure and character of the town derives from the industrial activities and their location. The starting point of interpretation is not the contrasted character of social areas but the location of industry.

Coal mining and urban settlements

The dominance by industrial needs was perhaps most marked in the coal-mining settlements where the priority of the needs of the pit and all the ancillary operations, including transport, relegated house building to the sites which were left. One small example is a fair indicator of the process. Jeffery Davies (1976) has been able to identify the background negotiations to the proposal to build the Rhondda village of Bryn Wyndham, or Tynewydd as it became known (Figure 6.8),

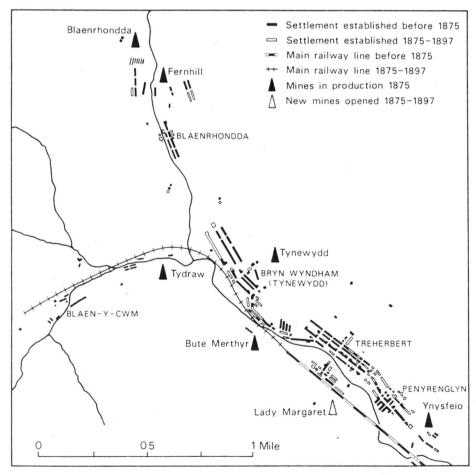

Figure 6.8 A The settlement pattern in the upper Rhondda Fawr 1875–1897. (after P. N. Jones, 1969, 78, Figure 11)

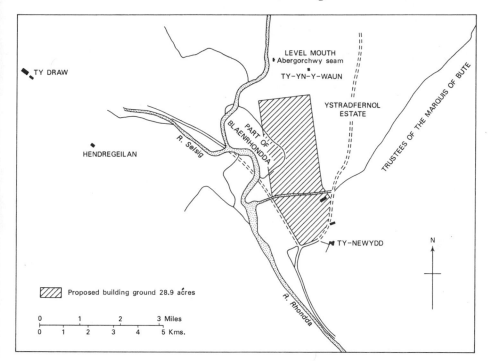

Figure 6.8 B Bryn Wyndham in 1858. (after J. R. Davies, 1976, 61, Figure 6)

from correspondence between Thomas Joseph, a colliery owner and one John Randall, the agent of the Countess of Dunraven, who owned the land on which Joseph wished to build. There is ample evidence that Joseph wanted to accommodate his workmen in a planned settlement. His first proposal was virtually in the form of a model village, although that notion was never explicitly used. It included houses of different quality graded into first, second and third classes. There was, however, disagreement with the Dunraven estate over the size of individual plots and the associated rents, Joseph wishing to keep plots small and rents low. He threatened to build elsewhere, on land belonging to the Bute estate. He eventually obtained agreement from Randall however, and subsequently commented, 'it will be necessary to change the building site, from the place which I first suggested to another equally suitable so as not to interfere prejudicially with the working of the mines etc.' This second choice (Figure 6.8B) was 'Ystradfernol farm at the boundary of that farm with the Bute-owned Tynewydd farm, and in the fork of the parish road' (Davies, 1976).

There are three aspects of this venture which are relevant. The first is that Joseph certainly considered the advantages of site, recording that 'it lies well, has good aspect and with good approaches to it'. But that came well behind the needs of mining as the reason given above – 'so as not to interfere prejudicially with the working of the mines' – indicates. That same order of priorities is apparent in the second choice – 'it will not be required by me for sinking upon, or as Tipping Ground'. The third aspect is a minor one relating to the link between mine and settlement. Superficially, it might seem clear the Tynewydd was a development as a settlement for Tynewydd Colliery (Figure 6.8A). The reality was more complex than it might appear from direct map interpretation.

Planned industrial settlement

There was manifestly an element of planning in the creation of Tynewydd, but to a degree, most new settlements were 'planned' in the sense that some attention was given initially to the layout of residential areas. But always, and even in much larger settlement, the demands of industry were paramount and of themselves created many of the problems the cities faced. Perhaps the town which attracted most attention during the second half of the nineteenth century was Middlesbrough. Few of its historians fail to refer to Gladstone's description of it as, 'This remarkable place, the youngest child of England's Enterprise. It is an infant gentleman, but it is an infant Hercules' (Gladstone, 1862). And as two commentators continue after introducing part of that quotation, 'The Victorians in general with their amazing capacity for self-congratulation, rhapsodized about this awesome offspring of their generation. Its neighbourhood, "rich as the mines of Golconda in subterranean wealth", the giant's growth according to its own official guide, "was unregulated, haphazard and aggressively utilitarian". For visitors to England in the later nineteenth century Middlesbrough was a more important sight than any cathedral: most nations had cathedrals, but only England had this monstrous vision of the industrial future' (Bell, 1972, 173).

Figure 6.9A and B shows land uses in Middlesbrough in 1853 and 1891–93. The town was the creation of Joseph Pease, who had been involved in the construction of the Stockton to Darlington railway which had begun operating in 1825. Pease, aware of the possibilities of greatly increasing coal exports, realized the limitations of Stockton and looked for a site nearer the mouth of the Tees. He chose land in the Chapelry of Middlesbrough, initially proposing the name of Port Darlington. Pease

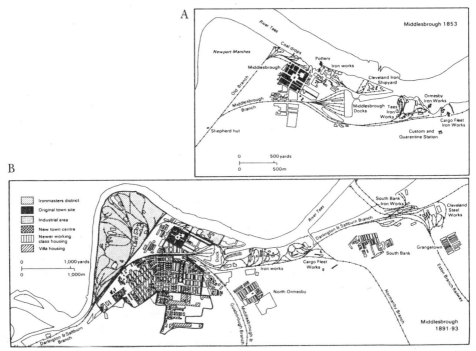

Figure 6.9   Middlesbrough in A 1853 and B 1891–93. (after Morris, 1968, 168, Figure 8 and 9)

formed a company and purchased 500 acres of the Middlesbrough estate in 1829. In 1801 the population had been only 25; it was 154 in 1831. By mid century it was to be 7,437 and in 1901, 91,302. The Bells' comment on this development is, 'Throughout its history the "great type of the iron age" has been dedicated to the view that a town is the place where people live, as uniformly as they may, in the trivial hours between the shifts' (C. and R. Bell, 1972, 193–94). That comment effectively encapsulates the dominant locational demands of industry. Residential areas were simply annexes to the industrial sites.

It is true that Middlesbrough was laid out carefully on a regular grid, although as one commentator notes, it 'would have suited an Irish plantation town' (Morris, 1986, 166). The grid of streets was located on the sole area of relatively dry land amidst the salt marshes. It was given a market in 1840, and a church. It was incorporated in 1853.

By 1840 the export of coal from Teeside had risen to one and a half million tons. A pottery had been started in 1834, but the major transformation came in 1850 with the discovery of reserves of ironstone in the Cleveland Hills. The man involved was one John Vaughan and in partnership with Henry Bolckow, he initiated the growth of the iron industry, buying six acres of the river bank in 1841. 'The region was completely transformed in appearance as a result of these economic advances. In 1851, the year of the Great Exhibition, the first blast furnace in Middlesbrough was blown in: before 10 years had elapsed there were 40 furnaces in Cleveland and Teeside' (Briggs, 1963, 256).

Now the critical issue here is not the general development of the town and the nature of its residential areas (Bell, 1972) or the general significance of Middlesbrough as a Victorian city (Briggs, 1963), but the nature of its internal structure (Morris, 1986). The river bank land, initially dedicated to coal export, became an extended zone of industry with adjacent residential developments, the annexes of industrial sites. Thus 'Bernhard Samuelson, son of a Liverpool merchant and head of an agricultural engineering works at Banbury, who, after building furnaces at Southbank in 1804, leased a 10-acre field at the rent of £5 an acre and set about creating yet another community of his own' (Briggs, 1963, 257). These works and residential adjunct can be identified on Figure 6.8B. There are very close parallels with Neath and Swansea, the same linear extension along a river, the same creation of a set of urban villages in relation to a commercial core, which itself was subject to pressures both by the presence of industry and by the forces of growth. Thus at Middlesbrough the boundary of the railway, which confined the planned core between it and the industrial river strip, led inevitably to detached growth 'the other side of the tracks' and the eventual deterioration of the erstwhile centre.

There were other problems at Middlesbrough. The site on river marshland meant that drainage was a constant problem during the century. Added to the damp was the massive pollution created by the industrial belt. In spite of some initial altruism which characterized the layout, the overwhelming demands of industry created the town and determined much of its nature. If Middlesbrough in general epitomizes 'the philosophy of aggressive industrialization' (Bell, 1972, 190), the analysis of its internal structure must begin where the town began, with the nature of the utilization of the land along the river front.

The railways and the city

Close associates of industry in their demands for urban land were the railways. Within urban areas their demands were considerable for they were not only made up of the permanent way and stations but also of goods depots and cartage yards, marshalling yards and sidings, repair shops and engine sheds. An account of 'the impact of Railways upon Victorian cities' must turn to Kellett's book bearing that title (Kellett, 1969). There the estimate of the proportion of land in railway use in central city areas in 1900 ranges between 5.5 and 9.0 per cent, the latter figure at Liverpool where the provision of port facilities increased the area so used. These were quite substantial sections. But the major phase of railway development had begun in the 1840s and lasted until the 1860s by which date the bulk of the system as it was at the end of the century had been constructed. But that period of peak development came at least one hundred years after the conventional date for the onset of the industrial revolution. In consequence the railway was a late-coming competitor for land in the city centre. Like most users it needed maximum access for the greatest profitability. But as Kellett notes in relation to Manchester, the city's 'primary function, evolving even before the railways were built, was as a market, banking, business and distributing centre. This central land-use was not merely prior in time to the railways it also preceded them, logically, in importance' (Kellett, 1969, 266). The clear arbiter in the railways' bidding for land was land-values. Thus in the City of London, dealers on the Exchange were able to outbid all comers. 'Land values . . . were six to eight times as high as even the most elegant West End residential addresses. They were even two or three times as great (for the smaller sites required) as the prices the railways were able to bid for central land. The core of the central business district, therefore, must be taken as immovable by direct railway pressure' (Kellett, 1969, 298–9). Under these circumstances the railways had to be content with encircling the emergent central areas of cities and responding to change rather than initiating it.

Dennis has stressed the same point and extended the discussion of the basic problems which faced the railways. 'Station sites were determined by the ease and cost of land acquisition, critical as companies were anxious to avoid delays in commencing operations' (Dennis, 1984, 127). Once stations were opened the inflation of land values generated new problems derived either from the new attractiveness of land or the 'ransom' propensity held by landowners. The railways 'could not contemplate buying out commercial interests, especially where there were complications of fragmented land ownership, multiple rights and demands for compensation for loss of "goodwill" . . . Legal expenses were minimized and negotiations least protracted if agreement had to be reached with only a few landowners, ideally institutional landlords such as charities, hospitals, schools and canals or aristocratic estates' (Dennis, 1984, 127).

The end result of all these problems with land values and land acquisition was to minimize the impact which the railways had upon the city centre in spite of their areal extent and visible significance, in reality and on the map. Kellett's conclusion, that the railways exerted only a limited influence on the emergent land-use pattern of the major cities, has been widely quoted and nowhere refuted. He does allow the one exception of Liverpool, although even there the heart of the business district remained fixed in its old location inland from the Pier Head and Castle Street. It was retailing which was pulled northwards along Lord Street–Bold Street, but with an extension in Renshaw Street and Lime Street (Figure 6.10). Even then, the area around Lime Street is named a 'hospitality' area following Ramsay Muir's nomenclature of 1910 (Muir, 1910, 304). That was most certainly apt as hotels and restaurants gathered about railway stations. The 'Railway Hotel', ranging from the largest and most luxurious to the much more modest, was one of the distinctive architectural features of Victorian cities.

It can be argued, however, that analysis based on a small sample of the largest

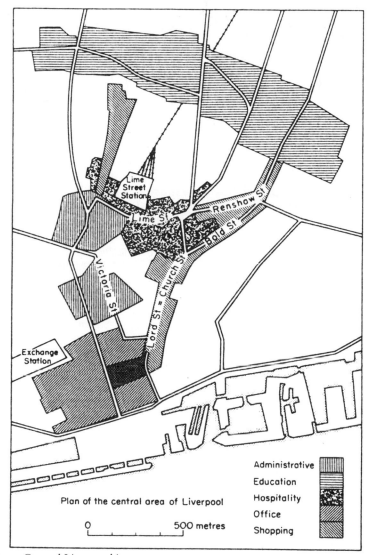

Figure 6.10 Central Liverpool in 1910.
(after Kellett, 1969)

cities is not a completely fair reflection on what happened in smaller towns. As Dennis writes, 'railways were more likely to affect town planning where business facilities were inadequate: expansion had to occur somewhere and the coming of the railway acted as both catalyst and magnet' (Dennis, 1984, 129). He cites the example of Huddersfield which the railway reached in 1847. The station, characteristically a building of some distinction, was built in an undeveloped area which had been acquired by the Ramsden estate. The initial idea was to lay out the area in the form of a new town but ultimately a grid-like plan was adopted. A large public square, St George's Square, was developed in front of the station and it grew into a significant part of the business sector of the town.

Reference can also be made back to an example used in the discussion of industrial areas. At Neath (Figure 6.5) the South Wales Railway was an integral part of the line of warehousing and factories which followed the line of the river. The station was sited well to the south of the medieval core. To a degree it was a response to the direction of town extension and that of the CBD, but it was also a stimulus to that movement. Again it was fronted by a square and across from it business was represented by banks and offices as well as shops. In an even smaller town, Aberystwyth, which developed as a resort, the railway station site was significant enough to play a considerable role in the creation of a completely new shopping area in the town, well outside the lines of the medieval walls (Carter, 1965). There seems little doubt that the land values were enhanced by station access and it was via that mechanism that impact was made and land uses influenced.

Whatever reservations may be held as to the role of the railway in the forming of the central business districts of towns, there is no doubt as to where its impact was most significant. That was upon inner-city residential areas, and especially those lowest in social status. Three associated aspects can be isolated. The first was that in terms of land acquisition, working-class areas were the easiest to obtain. Costs were lower and land easily assembled if it belonged to a single estate. Dennis catalogues the substantial removals of housing and populations to facilitate railway construction at Newcastle (800 families), Liverpool (500 families and later 135 houses), Sheffield (over 5,000 residents) and Manchester (312 houses). These details are available after 1853 when 'Shaftesbury's Standing Order' required 'Demolition Statements' which indicated the extent of working-class evictions as a consequence of railway construction. These were the data source used by Dyos in his earliest published work on the nineteenth-century city (Dyos, 1955, 2957), relating to London where the major displacements took place. These demolition statements, however, as Kellett points out, grossly underestimate the actual situation. Railway companies' returns were not checked; projects could be broken down into a series of small sections so that each one, individually, did not seem excessive; purchases of land could be made in advance so that unoccupied houses did not appear and with the added advantage that land was cheaper before the railways actually appeared in the market as purchasers (Kellett, 1969, 328–9).

The total area of central land taken by the railways during the nineteenth century was about 800 acres. Dyos estimated the population removed as 76,000 between 1853 and 1901. Kellett, for the reasons given, regards these, as Dyos indicated, as an absolute minimum and suggests something like 120,000 between 1840 and 1900, including the period before the Demolition Statement was required.

Geographically the consequence of this process was threefold. The displaced tenants had no right to rehousing; those who lost accommodation had to find it elsewhere, usually in the same vicinity. The result was a tendency to increase population densities and to exacerbate the slum conditions which already obtained. The second was to accentuate trends towards deterioration. If the railway section had a beneficial effect on land values, all the other aspects of railway land-use, the sheds, sidings and marshalling yards, had the opposite effect. Mixed with the railway uses and the warehouses were dwellings of the poorest category. The third was to introduce clear dividing lines into the city, distinctive barriers which in many cases cut off the poorest sections. Again reference can be made to Neath. The Green, marked on Figure 6.5, was cut off from the rest of the town by the railway and approached by a tunnel under it. The other boundary was an older transportation line, the Neath Canal. Thus isolated, it was literally on the wrong side of the tracks and one of the poorest areas in Neath. But railways acted as barriers on a much larger scale; the whole city centre of Cardiff was defined by the

South Wales Railway and the Taff Vale Railway, and has remained so until the present, so effective have the tracks been as limits.

Conclusion

A summary of the role of industrial and railway installations in nineteenth-century towns must surely begin with their physical dominance. It is usually represented in the symbolism of buildings. Castle and church dominated the townscape of medieval times, the tower blocks of commerce, finance and bureaucracy now define the skylines of cities; in the nineteenth century mill, factory, mine and warehouse stood out, along with gothic railway station and an associated hotel. But there was more to it than that. When furnaces were tapped the glare lit up the sky so their presence was reasserted daily. There was no pollution control so that the smell of the works permeated the whole region; the stink of canals and rivers into which effluent drained was an ever-present background. The pea-souper fog, if largely the product of domestic fires, had an industrial component to add to its density. The squalor of residential areas has pre-empted academic analysis, in reality the physical presence of industry and its associated land-uses were probably the most visible and tangible aspect of the city in the nineteenth century.

There are three aspects of industry which have emerged from this review as significant to the spatial structure of towns before the First World War.

The first is that small-scale workshops remained a significant element. In general, they were under threat from large-scale regionally specialized industry. Thus the bespoke tailor and boot and shoe maker were declining before the mass production of the off-the-peg item of clothing and footwear. But local production continued, the dressmaker well into the present century. Contracting out, sweating, still survived as part of the organized structure of industry. Its location became more and more specifically associated with that mixed zone of uses which surrounded the CBD. Obviously banished from the centre by rising land values, there were still possibilities in the back streets and courts of the earliest industrial-age extensions of cities. Burgess, in 1925, was to call it a zone of transition, abandoned by higher quality residence, outside the immediate demands of the city centre users of land, associated with railheads. In the biggest cities high specialization, especially on skill and value as in the gunsmiths of Birmingham, ensured survival as against the mass market. The inner city, with its additional mix of breweries, smithies, slaughter houses and such like uses, was therefore, until the century's end a significant industrial area.

The developments just outlined can be viewed as an intensification of past conditions. It is the second aspect in this summary which was novel, and that was the factory, the central symbol of industrial capitalism. But that is not the point as far as the spatial pattern was concerned. What was significant to the cities was the creation of greatly extensive areas dedicated to industrial and transportation uses. They were the creators of the towns, the first in the field so that as far as industry was concerned land acquisition was rarely a problem, it was clearly set aside, as at Middlesbrough. True, there were problems which occurred later with extension and with the later railway building, as has been indicated. That extension was often the consequence of forward and backward linkages to build up great complexes. The metallurgical industries spawned chemicals by the use of by-products. The demands of textiles for machinery created engineering industries. The prime need for effective communications for the assembly of raw materials and the distribution of the finished products necessitated a close association with the canals and railways so that all were locationally linked. It is interesting to compare this contrast of the inner city workshops and the extensive areas of factory industry with the ideal town set out by James Silk Buckingham in 1849. His town was made up of a series of nine squares. Manufacturing industry was located beyond the outer square, it was peripheral, in order to reduce pollution. The outer square was to consist of working-class housing, so retaining the propinquity of work place and place of residence. Next in from the working class houses was a square devoted to crafts and workshops. Buckingham thus retained the clear contrast between these two types of 'industry' but totally inverted their actual location in the city. What he did, of course, was to graft these activities on to a pre-industrial city where those highest in the social scale lived at the centre, producing an amalgam of early and modern town. But it is significant that the clear distinction was made between workshops and factory industry and the working class was located between the two. Move these inverted squares to the centre, and the basic structure of the nineteenth century city appears.

If industry was so dominant it raises the question with which this chapter began. Why has it been so neglected, why do not studies of the great nineteenth-century towns begin with industry? For example, a standard work, *The structure of nineteenth-century cities* edited by Lawton and Pooley, virtually makes no mention of industry as such, the three sections into which it is divided dealing with housing, retailing and social structure. To a degree the answer lies in academic fashion and the availability of a data source, namely the enumerators' books of the censuses, which has directed attention to social geography. But there is a more fundamental reason. Industry was not a significant bidder for city centre land. Once the business and commerce it engendered were under way they, in any case needing smaller areas, became the prime bidders and it was the prime bidders who determined the form of the city centre and so have attracted attention. Even so, the location of industry was still a major factor in the structuring of cities.

Nowhere was that more important than in the generation of contrasted residential areas, the third aspect of this summary. Early entrepreneurs lived in close proximity to their mills and factories. An article in the *Bristol Times* in 1851 comments on Sir John Guest's house in Dowlais (see Figure 6.7). 'His residence at Dowlais is a large, smoked, sooty-looking dwelling, overlooking the works . . . Our Lady Lindsay, his mother-in-law, in her hyperborean haughtiness, calls the house at Dowlais "the cinder-hole", but it is a cinder-hole from which her son-in-law can command more influence, and look down upon a far mightier, and to the thinking and philosophical mind, more sublime sight, and all of his own making, than ever the Lindsays could command from their castle.' But Guest owned a house in Spring Gardens in London, and in 1846 had bought Canford Manor in Dorset. Entrepreneurs were seldom totally committed to such work-side locations and especially later in the century, as joint-stock companies were formed, the on-site operations were left to managers who preferred to live in the newly growing residential areas of high status. Industry became a repeller of good quality housing. In contrast, when the journey to work was made on foot then the workforce of necessity became clearly linked with workplace. Entrepreneurs built housing for workers, greatly varied in quality, as part of their industrial development; in many cases in isolated areas it was a necessity, a factor of production. It was in this way that the urban villages which have been noted earlier came into being. Adjacent to industry, they were often circumscribed by railways. No doubt the isolation which resulted contributed to the development of community, but not to quality or to visibility. The bond between place of work and residence was not broken by the end of the century in spite of the growth of workmen's trains and it was probably in that context that industry made its greatest impact upon other land-users and general urban structure.

7

Demands for urban space: commerce and government

It would be a neat encapsulation of nineteenth-century changes of land-use in the city centre if they could be reduced to a fundamental transformation from an undifferentiated condition at the beginning of the century to a complex mosaic of specialized uses dominating discrete areas at the end. If the period be extended to take in the whole of the industrial period from the mid eighteenth century to the outbreak of the First World War then that summary might well be acceptable for the vast majority of towns, perhaps for all bar London. But within the overall changes there was a two-fold process of evolution. In the first case each of the relevant users, such as retailing or government, itself went through a process of metamorphosis. Thus, for example, urban government at the outset was character-istically confined to nothing more than a small town-hall. By the end, and especially after the reform of local government in the 1880s, there were large and elaborate city-halls and shire-halls as a response to county government, and a panoply of other buildings, perhaps gathered together to form a distinctive administrative area, or scattered throughout the centre of the town on a purely ad hoc basis derived from the availability and purchase of property.

The second aspect of the two-fold process was the progressive segregation of land-use into specialized regions, a condition which reached its farthest extent in the largest settlements where demands for space were greatest. In crudest terms, in the early eighteenth century retailing in the form of a weekly market might well have been located under or adjacent to the town hall. Its earliest location had probably been in the churchyard. By the end of the century retailing had become concentrated in distinctive shopping areas and along with administration, and indeed religion, occupied separate and distinct areas. It is difficult to follow these two processes, closely associated though they are, in tandem. In the first instance, therefore, patterns of change in each use will be followed separately, and then an attempt will be made to put them together as they determine the character of central areas.

Retailing and the development of the shopping centre

It is appropriate to begin with retailing, or, in more spatial terms, with the shopping centre since, with only some exceptions in the largest of cities, it was the use able to bid the highest for land and therefore, pre-empt the most accessible locations. That, at least, is the assumption behind the standard graphs of bid rents of competing

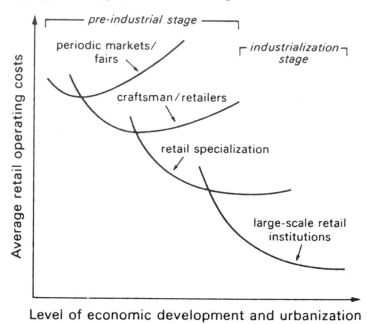

Figure 7.1 Stages of retail development. (after Shaw and Wild, 1979, 279, Figure 1)

users against distance from the city centre, and it is an accepted, if debatable tenet that competing bids for land controlled use in the nineteenth century.

Figure 7.1 represents a scheme set out by Bucklin in 1972 which plots average retail costs against the level of economic development and urbanization. Where the purchasing power of the population is low and diffuse then the retail system will be dominated by periodic markets and fairs. With economic development, purchasing power increases and can be concentrated so that the operating costs of the fixed dealer fall and the market becomes permanent, at first in structural terms in a market hall or such building, eventually in operational terms, that is shops open throughout the week rather than on a specific day or days. Craftsmen-retailers can now establish fixed shops. With further economic development the production of consumer goods becomes larger in scale and more standardized; it also becomes concentrated in favoured locations. The small-scale craftsman is thus undercut, though survival is possible by concentration on quality and the meeting of individual demands. Even so, the specialist retailer, whose expertise is in retail business rather than in the product, becomes dominant, whilst further increases in scale bring large-scale retail institutions, the department store for example, into being. All this is necessarily based on increases in both personal mobility and the nationwide movement of pre-packaged, standardized products. There is no doubt that Figure 7.1 fairly represents the process of change which took place in retailing, although perhaps a little greater allowance should be made for overlap; periodic markets have certainly not disappeared from the retail scene in Britain, although they now tend to be held on land away from city centres in order to reduce operating costs. What, however, is much more difficult is to relate the points at which the graph lines intersect, the time at which one form of retailing succeeds another as the dominant, to specific historical periods rather than refer to the very vague concept of level of economic development and urbanization.

There are a number of histories of retailing, some dealing with the topic in general, of which the best known is Davis's *A history of shopping* (1966), but most dealing with specific periods (Adburgham, 1964; Alexander, 1970; Fraser, 1981; Jefferys, 1954; Mitchell, 1984; Mui and Mui, 1989) or with special aspects (MacKeith, 1985). Unfortunately there is little in the way of clear consensus as to when the critical transformations took place, a problem which is made all the more difficult since they occurred at different times in different places, London was clearly the leader, and there was also a social class difference in shopping behaviour.

Of all the writers Adburgham is the most decisive in arguing for a major transformation pivoted about the end of the Napoleonic Wars. Before that point shop, workshop and home were all combined in one, that is, craftsmen-retailers produced goods for sale on the same premises where they lived, often with a number of apprentices. The skills handed on to the apprentices were concerned with the product, not with retailing as a business. Thus the tea merchant bought supplies in bulk from London and was an expert on teas and not someone selling a ready-packed standard product. The cabinet maker constructed and sold furniture usually made to a specific order, he did not simply sell factory manufactured goods. Prices were not fixed, there were no tickets indicating the cost of the goods and 'abatement' could be negotiated; in short prices were determined by haggling. The streets were unpaved and unlit so that 'shopping' as such, in the modern sense, did not exist; rather marketing was carried out by those lowest in the social scale, and where high quality goods were concerned, they were brought to the house for inspection. Shops did not advertise, although there were signs for the illiterate. The weekly market and the itinerant trader were the basic mainstays of the system. After the 1820s all these conditions were reversed. Permanent shops appeared and improvements in glass production meant that they could have lighted windows for display. Specialist non-resident retailers sold goods which were nationally distributed under brand or company names. These goods were offered at fixed prices and there was no abatement. Streets were paved and lighted so that visits to large, well-stocked shops became a pleasure and, indeed, a pastime for the well-off. Restaurants and tea-shops came into being to give the opportunity for rest and further to enhance the enjoyment of shopping. Shops began to advertise themselves and the national producers advertised nationwide the standard goods they were distributing. A wholesale system came into being as an intermediary between producer and retailer. The old weekly market became no more than a possibly busier day for shops, which were open all week, and in many cases 'all hours'. Annual fairs lost all their 'selling' function and became devoted to pleasure. It is true that such a transformation did not affect all types of shops uniformly. Perishable foodstuffs, such as fruit and vegetables, continued to be offered mainly in the market halls, greengrocery was one of the last operations to be located in fixed shops.

The description of changes in retailing which have just been set out certainly epitomizes the transformation which reached a climax in the first part of the nineteenth century, but to ascribe it to one decade, even the year of 1825 as Adburgham does, is clearly inapposite. Davis even finds the point of inflection very much earlier. 'If a critical point can be discerned in the slow evolution of the retail trade, then its date is the epoch of Elizabeth I and the early Stuarts; its place, London . . . Retailing began to come of age and to take seriously the business of wooing the consumer. Something that we nowadays recognize as shopping had begun' (Davis, 1966, 55). London was highly exceptional, but even so, and, referring back to Figure 7.1, Davis is identifying the break between the periodic market and the craftsman-retailer rather than the coming of modern shopping; and the weekly markets and annual fairs continued to be widely significant across the country. In their book *Shops and shopkeeping in eighteenth-century England*, H. C. and L. M. Mui make a much more thorough case for bringing back the transformation into the eighteenth century. If the onset of industrialization is taken as about the middle of the eighteenth century that is only to be expected for changes do not neatly relate to rounded dates, and the Muis may be describing the onset of a process which was completely accomplished by the first quarter of the nineteenth century.

Certainly 'The birth of a consumer society' is vigorously advanced in the book of that title by McKendrick, Brewer and Plumb which bears the sub-title 'The commercialization of eighteenth-century England' (1982). The book begins with the unequivocal assertion that 'there was a consumer boom in England in the eighteenth century. In the third quarter of the century that boom reached revolutionary proportions. Men, and in particular women, bought as never before . . . In fact, the later eighteenth century saw such a convulsion of getting and spending, such an eruption of new prosperity, and such an explosion of new production and marketing techniques, that a greater proportion of the population than in any previous society in human history was able to enjoy the pleasures of buying consumer goods. They bought not only necessities, but decencies, and even luxuries . . . For the consumer revolution was the necessary analogue to the industrial revolution, the equation to match the convulsion on the supply side' (McKendrick *et al.*, 1982, 9). However, some caution is needed and although it is presumptuous to accuse distinguished historians of being ahistorical there is a danger of transferring back into the past a totally inappropriate modern term – the consumer society. And any consideration of the lives of the urban poor – even those of the average working-class families or the greater proportion of the population – must give some considerable hesitation over accepting the idea of a consumer revolution, even at the *end* of the nineteenth century.

Such views are not without relevance to the discussion in Chapter 4 of the changing city system, at least to a degree they would support the significance of early mercantile activity and suggest that the emerging shopping system was a stimulant of industrial change rather than no more than a reaction to it. Extensive discussions of the number of shops, and the population per shop can be found in Alexander (1970) and the Muis (1989) but the data are so tenuous as to make any conclusion unreliable, but all the evidence quoted suggests a revolution beginning in the eighteenth century and reaching fruition in the first part of the nineteenth.

Against this discussion of early change can be set the rather greater amount on the late nineteenth century (Fraser, 1981; Jeffreys, 1954; Alexander, 1970). There the modifications are much more evident and agreed. Behind all the detail they were threefold. The first was the development of chains of stores, the multiples. Hamish Fraser estimates that in 1880 there were 14 firms with 108 branches in the grocery and provisions trade. By 1900 the figures had increased to 114 and 3,444 (Fraser, 1981, 116). Other characteristic multiples were in footwear with some 64 firms and 2,589 branches by 1900, and in clothing. By 1900 names such as Liptons, W. H. Smith, Freeman, Hardy and Willis and Boots were widely known throughout England and Wales. Jesse Boot was characteristic. Initially a grocer and herbalist, he began selling proprietary medicines in 1874. In 1886 he converted to limited liability to outflank the restrictions of the Pharmaceutical Society and expanded with branches throughout the Midlands. By 1900 there were 181 branches and by 1914, partly by take-over, there were 560. He undercut the small retailer in price and advertised nationally. When problems of supply arose he started his own manufacturing provision. He also diversified into books, stationery, fancy goods and photographic equipment (Fraser, 1981, 120; Chapman, 1974).

The second aspect of nineteenth-century retail development was the co-operatives. The well-known Rochdale Pioneers opened their first shop in 1844; the Co-operative Society was founded in 1863 when membership of Co-ops was about 100,000; by 1893 it was 1,680,000. It was significantly an attempt to include the working class within the new consumerism and to revolutionize the pattern of shopping for those to whom the standard developments in shopping were to a degree irrelevant. There were other co-operatives of which perhaps the Army and Navy Co-operative Society of 1872 is the best known.

The third feature of change in the late nineteenth century, usually dated after 1860, was the department store. The basis of the department store can be found much earlier in the century in the gradual extension of the range of goods sold by drapers and haberdashers. Mostly this was into areas where the female buyers of clothes were interested, into accessories and into carpets and household furnishings and equipment. The well-known names of the London stores which marked the peak of that type of retailing – Debenham and Freebody, Bourne and Hollingsworth, Peter Robinson, as well as Lewis in Liverpool and Manchester, began in that way. Closely allied was the growth of the seller of a variety of products in weekly markets, such as Swan and Edgar and the most well-known of all, Marks and Spencer. Michael Marks was a penniless Russian immigrant who arrived at Stockton in 1881. He set himself up as a pedlar in Leeds with a tray of goods and because he spoke no English, the famous sign on it read 'Don't ask the price, it's a penny'. By 1884 he was renting a stall in Kirkgate Market. From thence he moved into the covered market and by 1894 he was operating eight penny bazaars. By 1900, after taking Tom Spencer as a partner, Marks and Spencer were operating in 24 markets and 12 shops. But the epitome of the department store was that opened by Gordon Selfridge in 1909. Sales were promoted by elaborate visual display and by an environment which was warm, bright and glamorous. Closely related to the department store, but still somewhat different, was the variety chain store which sacrificed atmosphere for a functional cheapness and aimed at a lower class of purchaser. Woolworths is the outstanding example, the first store in Britain opening in Liverpool in 1909.

A summary of these retail developments during the nineteenth century suggests a highly complex process with a number of distinctive characteristics.

1   There was an extended period of evolution, virtually from the sixteenth century, which saw the replacement of weekly market and annual fair by permanent specialized shops open throughout the week. But the pace of change quickened throughout so that the critical feature, the emergence of a true shopping centre, as a standard urban feature, can be seen as being achieved during the first part of the nineteenth century. Paved and well-lit and with well-lit shops having display windows, it was a place which provided a leisure activity as well as performing a retail function. This rapid pace of change was closely associated with the increasing affluence of the middle classes and the onset of conspicuous consumption.

2   Late in the century, mainly in the last quarter, the development of chain stores, variety stores, co-operatives and department stores marks the onset of mass consumption and the transformation of the city centre by the growth of uniformity derived from the nationally distributed shops and their standard frontages.

3   These changes did not take place across the country at the same time. There was

virtually a hierarchical procession of change beginning in London, then characterizing the largest towns and cities before filtering down to the small country towns.

4   The market hall can be regarded as an intermediate stage, a permanent structure as opposed to the temporary street stalls of the traditional market day, but still a step removed from permanent, detached shops. It is possible to argue that there was a class difference, that the specialized shops served the well-off while the working classes used the market halls. This was true to an extent, but there was also a functional contrast for perishable products remained the basis of the market halls virtually throughout the century, as already suggested the greengrocer was one of the last specialists to set up in shops. Whole series of market halls were established to meet demand and as Scola notes, 'Until 1846, butchers and fishmongers were forbidden by manorial law to open a shop in the township of Manchester outside the allotted market (Scola, 1975, 159–60). Some other relevant aspects can be added:

5   In order to transform the centre into a retailing area a formal and effective market in land was essential. If at an earlier period land was acquired mainly as a guarantor of social rank and prestige, by the nineteenth century land was a commodity bought or leased in anticipation of gain. At what date it is appropriate to argue that bid rents against distance became the arbiter of use is not easy to ascertain. But certainly during the nineteenth century urban land was treated as a commodity to be bought in the market. Urban land values, or their surrogate in rateable values, become relevant predictors of use. Some caution is needed, however, in the application of a bid-rent determinant of land-use in that urban land was still held by the large estates who could greatly modify straightforward economic processes.

6   The one further aspect of retailing which has not been discussed is the coming of the lock-up shop. Alexander notes that 'the lock-up shop was still unusual in 1850' (Alexander, 1970, 11). Indeed the traditional apprentice system, where living-in was an integral part, lasted on until the later part of the century. This was especially so in drapery and grocery, long after the basic need for specialization in the product had been greatly modified by ready-made clothes and pre-packaged food. It was therefore, in the later part of the century that the abandonment of apprentices, of living-in, and the adoption of the lock-up shop so that the owner escaped nightly to the suburbs, created the modern central business district.

7   Before the end of the century, and as a response both to the suburbanization of population and the transport systems which sustained it, a process of retail decentralization had set in (Shaw, 1978). Food shops, especially the independent grocers and butchers rather than the multiples, moved to the lower-order centres which were developing within the inner suburbs or to the characteristic linear extensions from the centre. Reference back to Figure 6.5 will illustrate the latter feature in Neath along Windsor Road. The result was that in the larger towns of England and Wales the density of shops increased until 1850 but then decreased as the smaller independents moved out and the larger multiples and chains occupied the more extensive central sites. Shaw has illustrated these developments in Hull where even by the first quarter of the nineteenth century small suburban centres had come into being although it was only in the second half that the combination of lower-order shopping centres, often linked by linear extensions had emerged (Figures 7.2 and 7.3) (Shaw, 1988, 240–5). Examination of the intra-urban hierarchy demonstrates that 'a diverse range of shopping centres existed within fairly close proximity to one another. For

example, in the city there were 38 local/neighbourhood shopping centres, located within a net built-up area of 771 hectares, with the greatest concentration occurring in predominantly working-class suburbs. In addition, more localized needs for basic groceries were provided for by over 500 shops located outside recognizable centres. The two 'second order' centres both located at short distances away from the central retail area, evolved from smaller shopping complexes to cater for a widening working-class demand after 1870' (Shaw, 1988, 243–4). There was, therefore, a significant contrast with what was to occur in the second half of the twentieth century. High quality shopping remained lodged in the centre, whilst lower-order shopping centres and corner shops met the localized demand from a working class with restricted mobility.

8 As demand for retail space, and land values, rose in city centres, the problem arose as to how to maximize shopping frontages and space about the peak land value. There was no way of building high before the era of the steel frame building and the lift. The response was the development of shopping arcades which were no more than pedestrian precincts providing both shelter from the weather and relief from the noise and dirt of city streets. It is not surprising that one of the first to be constructed was the Burlington Arcade which still remains as the longest, busiest and most prosperous (MacKeith, 1985). Margaret Mackeith in her gazeteer of arcades identifies some 73 constructed in or before 1900 which are still extant and 36 which have been demolished (MacKeith, 1985, appendix).

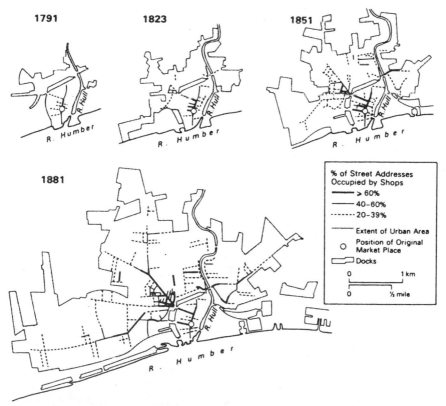

Figure 7.2 The development of shopping streets in Hull.
(after Shaw, 1988, 240, Figure 16.3)

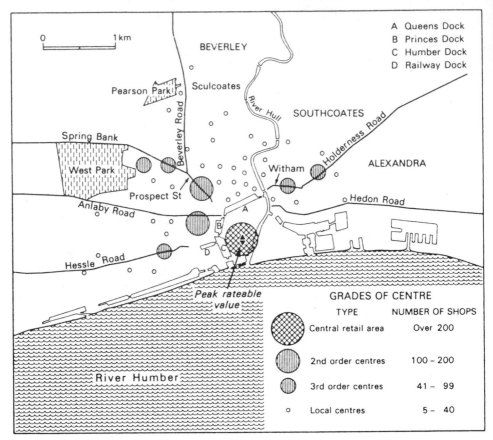

Figure 7.3 The intra-urban shopping hierarchy of nineteenth-century Hull. (after Shaw, 1988, 244, Figure 16.5)

The financial area and warehousing in the city centre

Retailing has been extensively considered in terms of its evolution during the nineteenth century since it was the predominant central use common to all towns. There is a whole series of other uses which sought central locations – financial and office, administration, entertainment, warehousing and storage – yet in very few cities was their extent sufficient to create distinctive areas. They either became merged within retail centres, as did banks for example, or pushed into the mixed transitional zone which surrounded the city centre, or into the upper floors of central buildings as was common for offices such as those of solicitors or insurance agencies.

What can be referred to in very general terms as finance and business was the major non-retailing element which in larger cities could compete with retailing in terms of bidding for land. Kellett identifies that situation in London. 'Because they set in motion a whole chain of operations in manufacturing, warehousing, transporting and selling, and because they depended to such a marked extent, even in the days of the telegraph and efficient postal service, upon the personal meeting, the dealers at the Exchange and their associated offices and credit agencies, were able to bid for central location against all comers. Land values in the City of

London were six to eight times as high as the most elegant West End residential addresses' (Kellett, 1969, 298–9). This ensured that the City of London remained as a predominant financial and business centre, itself becoming progressively characterized internally by highly specialized sections, whilst retailing was pushed westward, at the same time as it was being pulled in that direction by the westward spread of aristocratic and 'quality' residence.

In all this the big feature was the Stock Exchange itself. Although there was a securities market in London from the end of the seventeenth century, the Stock Exchange was not formally constituted until 1802 (Cottrell, 1986, 154). Stock markets were established in Liverpool and Manchester in the 1830s but after the boom and bust of the railway mania period there were only some nine exchanges outside London in the early 1850s (Morgan and Thomas 1969; Killick and Thomas, 1970). Thus only the largest cities had exchanges and the clustering of business offices which developed about them. In Liverpool the office area identified on the map of 1910 (Figure 6.10) and lying between the river front and the 'administrative platform' as Kellett called it, is a good example. But perhaps Manchester, the city par excellence of the business man, demonstrates most effectively the way in which financial uses developed early and subsequently

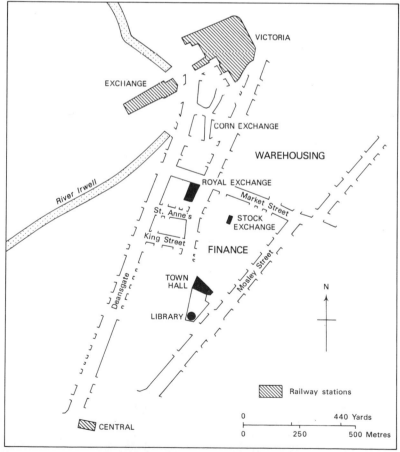

Figure 7.4 Manchester: The central area in the nineteenth century.

maintained a location over time. Manchester in the early nineteenth century has been described by Rodgers as 'little more than an overgrown market town not only in its size but in its structure. Its streets and squares were a chaotic jumble of buildings of different types and uses: houses, shops, workshops and even the earliest factories jostled each other in formless confusion' (Rodgers, 1961, 2). Perhaps that underestimates the eighteenth-century growth of the city which Briggs quotes a French visitor as calling 'a large and superb town' in 1784 and whose population had risen from about 40,000 in 1780 to over 70,000 in 1801 (Briggs, 1963, 85). Even so, the description of the organization of land use was probably apposite for it still displayed a lack of differentiation. Certainly high-quality residences were located at the centre. Market Street was the main axis and it was about it that the high-quality residences were distributed. In 1822 James Butterworth described St Ann's Square as 'the most opulent part', whilst King Street was termed 'elegant' (Butterworth, 1822). (See Figure 7.4.)

But the rapid, perhaps remarkable, development of the city in the middle of the first half of the century was to bring about a transformation. 'The population of Manchester in 1831 had increased nearly six times in 60 years, and by nearly 45 per cent in the previous decade, its greatest decennial rate of growth in the nineteenth century' (Briggs, 1963, 85). That explosive growth, together with the physical and socio-political character of the centre generated the first suburban exodus and the houses abandoned as residences became used as business offices or replaced by

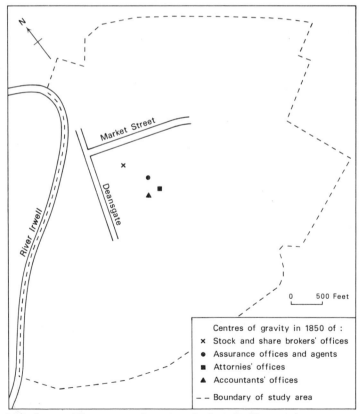

Figure 7.5 Manchester. The centres of gravity of selected office uses in 1850. (after Varley, 1968, Figures 33 to 36)

warehouses. 'By the 1830s the warehouses of Manchester were more impressive than the mills; massive, simple, austere, they were later to be praised for their real beauty . . . It is interesting to compare the 1839 and 1842 editions of Love and Barton's handbook to the town . . . "within the last few years Mosley Street," we read "contained only private dwelling houses: it is now a street of warehouses. The increasing business of the town is rapidly converting all the principal dwelling houses, centrally situated, into mercantile establishments, and is driving most of the respectable inhabitants into the suburbs"' (Briggs, 1963, 103).

Varley (1968) has mapped from trade directories the locations of stock and share brokers, accountants, assurance offices and agents and attorneys in 1850 and has calculated the centre of gravity of each. These are shown on Figure 7.5. These financial and associated uses clearly dominate the area south of Market Street and east of Deansgate, often utilizing the large houses from which the occupants had moved to the suburbs. It was closely associated with the Manchester Exchange, opened in 1809 and extended in 1838. This was the core of the Manchester merchants, wealthy and powerful enough to resist any challenge to their occupation. Varley also plotted the centres of gravity at three different periods and has demonstrated that throughout the century and into the second half of this, there was very little change indeed. As in London, the financial area established and maintained a distinctive location where the associated professions which needed face-to-face contact also collected.

Warehousing, too, as the development of Mosley Street suggested, took over former residential areas. Warehousing is often regarded as one of the dominant features of the Victorian townscape, 'massive, simple and austere' as Briggs described it, but there were two aspects of it, one dominating the early part of the century, the other the later. The growth of the textile industries during the eighteenth century and the production of cloth generated the large warehouses which made such an important contribution to the skyline of northern towns. Manchester especially was characterized by these buildings. Later in the century the development of wholesaling as an intermediary between producer and retailer generated a different demand for warehousing. Often the rears and the upper floors of retail outlets were used, another factor in the demise of the shop/residence, but more significantly depots were established usually at the margins of the central area and often near to railway stations. These later wholesale warehouses were of a different and more modest character compared to the industrial warehouses and, more diffuse in their location, made a less significant contribution to the city skyline. To return to the consideration of Manchester, in general Market Street acted as a divide with the major warehouse area developing to the north. Market Street itself had always been the main shopping centre and that role it maintained and developed into a more exclusive specialization, as did Deansgate which at the beginning of the century displayed a great variety of uses.

This section began with an emphasis on the financial areas of cities but, necessarily, it has become involved with warehousing and shopping. Indeed, a stage has been reached where it would be possible to propose a simple model of the way in which land uses and land users become segregated into distinctive parts of the city. But before moving to that more general discussion, it remains to consider one of the other demands for land which experienced major growth in the nineteenth century, and that was the administration which more and more was needed to ensure that the city continued to operate.

Administrative demands for city land

It has been contended that retailing and warehousing experienced a two-fold transformation during the longer period between 1750 and 1914. It is equally possible to propose a two-fold process of change in the administrative use of urban land. In the earlier period there was a whole series of buildings used as places of government and of assembly. They went under a variety of names – Moothall, Guildhall for example, together with Court House, Assembly Rooms and even Corn Exchange and Market Hall. The Municipal Corporations Act of 1835 introduced the first period of change. Incorporation brought with it the desire to express civic dignity and unity in a town hall. Many were relatively modest buildings, such as Hanson's original design for Birmingham in the 1830s. The second period of change gradually built up after 1850 and culminated in the Local Government Acts of the 1880s which created County Boroughs and the system of local government which was to last until 1974. The period was characterized by a vast increase in legislation which had to be carried out by local authorities. A great local bureaucracy was created and it had to be housed. Office space especially was needed, but also rooms appropriate to the dignity of the Mayor, and banqueting rooms to provide the proper environment to celebrate events of national and civic significance. A huge number of rooms was required 'to house the growing army of clerks. A glance at the growth of municipal legislation shows the extent of the explosion and by the last quarter of the nineteenth century town halls were needed to house town clerks, borough treasurers, borough surveyors and engineers, rating departments, gas, water and sewage offices, medical officers, market inspectors, nuisance inspectors, highways departments and the whole paraphernalia of school boards and even Poor Law guardians, not to mention providing adequate storage for documents. In architectural terms this might mean a plethora of separate buildings, but the complications of scattering public servants were often the excuse for building huge centralized town halls or civic centres' (Cunningham, 1981, 6).

The last sentence is particularly important. The demand for city centre land from other uses, the need to associate the various departments of civic government and the characteristic Victorian desire for the display of wealth and power, all led to that magnificent group of buildings, 'The Victorian and Edwardian Town Halls', so effectively illustrated by Colin Cunningham (1981) in his book of that title. As a result, although one of the most distinctive elements was added to the townscapes of nineteenth-century urban settlements, seldom did a very distinctive area or region as such emerge. What could not be fitted into the town hall was usually scattered about the town, again, very often in the transitional surround of the business area.

Two brief examples can be given. Middlesborough has already been considered in relation to individual land uses. Its first town hall was built in 1846, appropriately in the centre of the square of the planned town (see p. 83), a modest affair which still survives 'marooned in a wilderness of flat blocks' (Bell and Bell, 1972, 181). But expansion beyond the railway line and the creation of the new city centre there, as well as all the pressures of civic legislation, created the demand for a new town hall which was designed and built between 1883 and 1888. Figure 7.6 shows the ground floor plan. A great variety of activities were all housed together, including a chief constable's dwelling and a library! Since all these were separate, and some antipathetical, such as magistrates, prisoners and witnesses, separate entrances were provided virtually for each activity, a complex solution but simple as contrasted with the greater design problem of providing effective internal

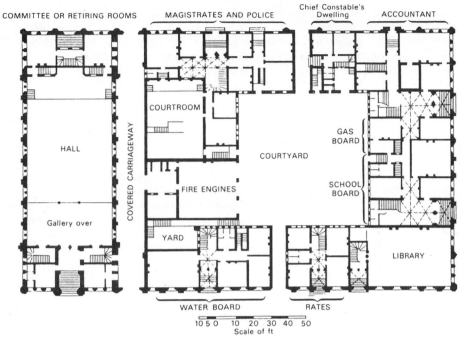

COMMITTEE OR RETIRING ROOMS   MAGISTRATES AND POLICE   Chief Constable's Dwelling   ACCOUNTANT

HALL

Gallery over

COVERED CARRIAGEWAY

COURTROOM

FIRE ENGINES

YARD

COURTYARD

GAS BOARD

SCHOOL BOARD

LIBRARY

WATER BOARD   RATES

10 5 0   10   20   30   40   50
Scale of ft

Figure 7.6 Middlesborough Town Hall.
(after Cunningham, 1981, 29, Figure 13)

circulation. The two town halls at Middlesborough neatly encapsulate the characteristics of the two phases of administrative growth during the nineteenth century. The second town hall clearly indicates how it pre-empted the creation of an administrative region within the city by becoming one itself and housing most of the essential departments.

Cardiff is one of the best examples of the forces at work behind, and the creation of, an extensive and elaborate administrative region. Cardiff was a late developer amongst British cities in the nineteenth century and it was not until the early years of the twentieth century that administrative demands became such that action became imperative. The old town hall was not only totally inadequate but road improvements demanded its removal. Existing administration was scattered over the city and so attention turned to the largest area of open land which constituted Cathays Park (Figure 7.7). It belonged to the Bute family, having been purchased by the first Marquis in 1793. The idea had been first mooted in the 1850s and in 1859 an approach was made to the Bute estate (Chappell, 1946, 13). It was not accepted, however, and it was not until after a series of abortive attempts that in 1887 some 58 acres were purchased by the city. The layout was, to a degree, determined by an elm-tree avenue which had been established by Lord Bute in 1878–90 which was to become the main axis of the new administrative region as King Edward VII Avenue. The actual plan of development had been established by 1903 and work began in that year. Sites were set aside for a new City Hall, a National Museum, for the University College which had been established in 1883, for a technical college and also for a University of Wales Registry. Associated was a series of formal gardens. Thus, there was established in Cathays Park one of the most distinctive areas of civic buildings in Britain. Of course, the added impetus of national (Welsh) buildings, as Cardiff's role as *de facto* and eventually *de jure*

(1955) capital of Wales developed, gave it a significant lift above other cities of a similar size. But Cardiff's Cathays Park is one of the first and most effective examples of the break from a single massive building to a distinctive urban administrative enclave within the city. It is also worth noting the distinctive role of the landowner and of the large Bute estate in Cardiff (see Chapter 9).

Other users of central city land

Up to this point a review has been undertaken of the various types of land use which competed for central city locations during the nineteenth century. Retailing, finance and business, commercial offices and warehousing were the main bidders, along with the railways and industry considered in Chapter 6 and residence to be discussed in Chapter 9. There were others. Entertainment is one, for example, but in even the largest cities it was made up of little more than a theatre or two together with the city hall, which was often the venue of concerts. 'The Messiah' sung at the local town hall was one of the standard experiences of the nineteenth-century urban mass, although occasionally, a Philharmonic Hall or even a Temperance Hall was a substitute location. But probably only London could have claimed anything which can be called a 'theatreland'. There were groupings of uses. Muir's map of Liverpool in 1910 which has been referred to on a number of occasions, identified a 'hospitality' area about Lime Street station based on the association of hotels which were attracted to the station, together with restaurants. Perhaps it also has overtones of less moral goings-on as well. Also in Liverpool during the century, derivative from the large merchant housing which characterized the Abercromby Ward, Rodney Street had acquired an exclusive medical role, analogous to Harley Street in London. Of the 88 properties recorded in Gore's Street Directory of 1906 some 77 were occupied by surgeons, physicians or dentists, some in multiple-occupation. It could well have been identified on the map as a medical area but it merged into the surrounding residential section and, indeed, in most cases there were mergings and interspersions of uses rather than a single use dominance.

The development of the city centre

The discussion of discrete use-areas poses the question of threshold populations, that is of the size and dates at which distinctive 'regions' crystallize out within the general central district. There are parallels with the definition of distinctive rank-levels within an urban hierarchy. But those levels are seldom easily identified and there have been few attempts to relate the segregation of uses to size at any other than a very general scale (Carter, 1972, 203). The notable exception is M. Bowden whose work, although taking in London, is mainly concerned with the USA and San Francisco in particular. The results Bowden calls tentative and certainly they reveal inconsistencies and, given the American context, are of little help. There are so many difficulties of definition and such a total inadequacy of empirical work that it is impossible to provide even the roughest of guides. Moreover, there is an additional dependence on the nature of function as well as size. Cardiff's highly distinctive administrative area at Cathays Park was a product of role rather than just of population numbers even if they be extended to take in the largest tributary area, the whole of Wales. In consequence, all it is possible to argue is that the separation and segregation of land uses and the creation of distinctive use regions was a function both of size and role. Only in London was a situation reached where full range of use areas had developed by the end of the century and where an additional process of highly specialized sub-division had taken place.

It follows from the above that the sequence of changes in the transformation of a partially differentiated core into a mosaic of specialized areas is also a topic on which but little has been written other than in general terms. Certainly there are few studies which trace detailed land-use changes over small intervals during the century. One such piece of research on Cardiff has examined changes in use in each building in selected areas marginal to the medieval core over intervals of 10 years or less between 1855 and 1949 (Morgan, 1985). At this point some brief account of the developing structure of Cardiff is necessary (Figure 7.7). The core of the city was the area contained within the medieval walls and that, because of the late development of the city (Chapter 4) remained functionally undifferentiated to any degree until well past mid century, hence the 1855 date at which the study referred to above begins. With the growth of coal export and a spectacular increase in population after 1860, the decade 1871 to 1881 was that of maximum percentage increase in Cardiff compared with 1821 to 1831 in Manchester noted earlier, the core was translated into a specialized retail and business area. The demand for floor

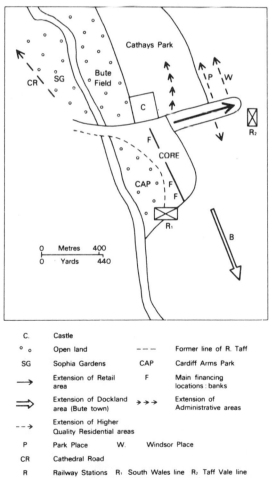

C.	Castle		
° °	Open land	— — —	Former line of R. Taff
SG	Sophia Gardens	CAP	Cardiff Arms Park
→	Extension of Retail area	F	Main financing locations : banks
⇒	Extension of Dockland area (Bute town)	⇢⇢⇢	Extension of Administrative areas
--→	Extension of Higher Quality Residential areas		
P	Park Place	W.	Windsor Place
CR	Cathedral Road		
R	Railway Stations R₁ South Wales line R₂ Taff Vale line		

Figure 7.7 The general structure of Cardiff in the late nineteenth century.

space was met within the core by the development of arcades, and externally by a clear eastward-directed expansion. To the west of the retail core a somewhat inchoate fringe of financial and wholesale buildings, probably related to the location of the pre-industrial quay, represented only in a very attenuated form the parallels in Manchester. The poor representation was due to the development of new docks well to the south, at the mouth of the River Taff. It was at that second nucleus that the main financial and warehousing facilities were located, including the Coal Exchange. Thus two of the significant competitors for central urban land were effectively abstracted into the completely separate nucleus of Bute Town. The open land of the Bute estate, later to become Cathays Park, has already been noted in the discussion of administrative uses. By the third quarter of the century the segregation of high-quality residence was incipient, producing two closely associated linear extensions in Park Place and Windsor Place, whilst there was also a minor extension southwards in Charles Street. Three of these critical areas (see Figure 7.7), the major eastward extension in Queen Street, one of the northern extensions in Park Place and the southern development about Charles Street, were examined by P. H. Morgan in the study to which reference has been made.

Tables 7.1, 7.2 and 7.3 indicate changes in crude land use in each of the areas at a series of dates between 1855 and 1910. The data are derived from a reconciliation of trade and street directories. Problems of reliability of such data have been discussed in Chapter 2, and it is necessary to treat the figures with some reserve.

Table 7.1 demonstrates the transformation of Queen Street and its progressive assimilation into the emergent CBD. In 1855 it was dominated by two uses, domestic or residential and the mixed category. It was at that date a primarily residential area, indeed it was traditionally one of high quality, representing the first push-out of the wealthy from the medieval core. But already by mid century it was showing a mixture of other uses.

The period between the directories of 1875 and 1882 saw a major change. Residence halved, and was to fall to a mere 6.7 per cent of uses recorded by 1890. Business uses, a category which includes shops and offices, increased from being negligible, other than within the mixed category, in 1855 to a quarter of the total in 1875 and then more than doubled to become the dominant use by 1880, and then virtually the exclusive use by 1890. A breakdown of the business uses, which Morgan also tabulates, shows them to be split fairly evenly in 1890, but with shopping taking the lead in 1904, only to fall back by 1910. But this was the characteristic situation of ground-floor shopping with upper floors being used as offices.

Table 7.2 presents the equivalent data for Park Place, the northern extension overlooking Cathays Park. There, domestic uses, mainly high-quality residence,

Table 7.1 Cardiff: Queen Street: general land-uses

	Vacant	Domestic	Business	Mixed	Education	Institutional
1855	0	34	1	56	2	1
1858	7	31	9	46	3	1
1875	5	29	24	44	1	1
1882	4	15	62	21	1	0
1890	5	8	104	1	1	0
1904	2	3	123	0	1	1
1910	2	3	124	0	1	2

Note: Institutional includes public and entertainment uses.

Table 7.2 Cardiff: Park Place: general land-uses

	Vacant	Domestic	Business	Mixed	Educational	Institutional
1855	0	7	0	3	1	0
1858	0	7	0	4	1	0
1875	0	26	0	0	0	0
1882	1	42	0	0	1	0
1890	4	36	2	0	2	0
1904	1	35	5	2	1	1
1910	0	28	14	2	2	4

remained dominant to the end of the century, strongly resisting the encroachment of business uses. But the latter had begun to intrude especially into the southern area adjacent to Queen Street. Park Place was broken by the docks feeder, which acted as a divide between the change to business uses to the south of it and the residential uses which remained to the north. By 1910 business uses, 83 per cent being offices, were developed enough to equal half of residential uses.

Table 7.3 shows the situation in the Charles Street area. The position in each of the areas is made more complex because of the additional premises added between each date. This is especially so in this area which was initially confined to Charles Street in 1855 but where other streets, Edward Street and Pembroke Terrace for example, were added, so that there is a great leap in the total number of premises between 1858 and 1875. Domestic uses remained the largest use until the end of the century but whilst Charles Street itself retained its middle to upper-class character the added streets were generally lower in social status. However, it also displayed a higher level of vacancy as development took place, and after 1880 there was a significant change with business uses increasing rapidly as domestic use declined. There was also a scattering of other uses as this became very much of a mixed area, indeed a transitional zone of the CBD. A separate analysis of the business uses shows that though offices dominated there was a characteristic wide spread, including wholesale premises as well as a small number of shops.

These three examples demonstrate the pattern of CBD expansion as residence was slowly pushed out and business took over. Specifically the extension of the shopping area, the emergence of an office area and the creation of part of a zone in transition are identified. Morgan represents each date by a diagram but only that for 1890 is reproduced as Figure 7.8. The various extensions of the CBD of Cardiff are clearly shown and the way in which the present mosaic of areas was created out of an undifferentiated core. It is atypical and unfortunate for demonstration purposes that industry, finance and warehousing were taken out by the growth of

Table 7.3 Cardiff: Charles Street Area: general land-uses

	Vacant	Domestic	Business	Mixed	Educational	Institutional
1855	0	29	3	10	0	3
1858	2	17	2	22	0	3
1875	7	151	8	15	1	5
1882	15	151	11	17	1	6
1890	4	154	14	23	1	9
1904	14	118	50	15	1	19
1910	11	102	69	11	1	23

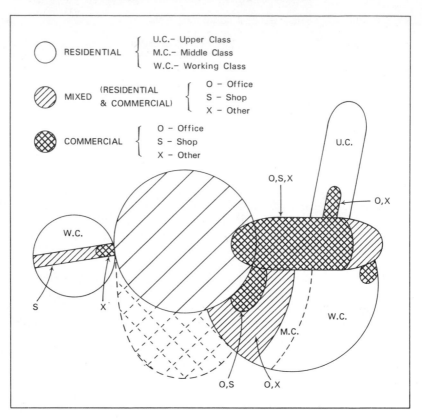

Figure 7.8 A generalized diagram of land-uses on the margins of the core of Cardiff in 1890. (after Morgan, 1985, Diagram 4)

Butetown about the docks. In compensation the distinctive administrative area has already been discussed (p. 101).

The diagram in Figure 7.8 has affinities with one of the few attempts to present a descriptive model of the emergence of the structure of the central area of the city in the nineteenth century (Figure 7.9), that by David Ward for the USA, derived mainly from a study of Boston (Ward, 1966, 1971). Ward writes, 'Until the nineteenth century, apart from the exclusive residential quarters of the rich, the functional specialization of urban land uses was only weakly developed. Most industrial and commercial activities were conducted on the premises of the producer or merchant, and local purchases or services were obtained on a custom basis directly from the producer' (Ward, 1971, 87). Accordingly, he identifies two stages of change, the first in the period 1840–70, the second from 1870 to 1900. In the first of these stages the dominating feature is the appearance of a distinct warehouse district, a consequence of increasing commercial activity. Storage space was in demand once a commodity or product was handled by a middleman between producer and retailer and once manufactured goods were distributed on a regional or national basis. Integral with such developments were the financial and insurance interests on which trade relied and hence a small but distinctive financial area appeared. Between 1840 and 1870 the main process was progressive specialization of these areas but after 1870 other uses were spun off. Amongst these was an administrative area as a response to the elaboration of urban government. Retail trade, which had tended to be scattered and interspersed amongst the finance and warehouse uses, became a predominant activity in a specialized area, a shopping centre appeared.

Ward's scheme is derived from the USA and if the development of land-use areas as discussed in this chapter are accepted then it would seem that for most British cities the timing would have to be moved back by several decades. Certainly the view that a specialized retail area can only be identified by 1900 hardly fits in with the notion of a consumer revolution in the eighteenth century, nor indeed with Adburgham's nomination of 1815 as the critical date at which modern shopping

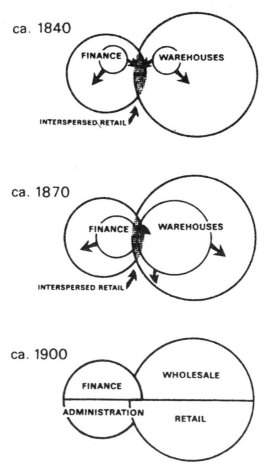

Figure 7.9 Generalized Stages in the Emergence of the CBD.
(after Ward, 1971, 88, Figure 3.1)

came into being. But the separation of financial uses and warehousing is very close to the reality at Manchester although it was achieved well before 1840 and the warehousing was not primarily related to retail trade. Again, the retail area, given all that has been presented earlier in this chapter, must be considered as an area in its own right rather than interspersed through other parts. Although financial uses can be ascribed dominance as the most successful maintainer of location, it would be best to regard all these uses as developing together in the largest cities but in overlapping association.

The other aspect which Ward's model ignores is the expulsion of industry and residence, the two uses least able to compete in bid-rent terms. Because of the amount of land needed, industry was least able to compete, other than in the location on upper floors of sweated trades or highly specialized and skilled activities. Thus in London silk weaving had been pushed out to Spitalfields and metal working to Clerkenwell, both to the north of the city, at an early date. As Chapter 5 demonstrated, industry occupied distinctive regions closely linked to transport. At Cardiff, the docks excavated to the south of the city abstracted industry. Residence, of course, remained in the general central area in the form of inner-city slums and ethnic enclaves, but that is a topic taken forward in Chapter 9. Certainly high-quality residence, which once occupied the centre, was expelled, or rather took off. Even in London it was pushed into the new estates of the West End. In most cities it was dispersed to the suburbs although, as Cardiff shows, it remained well beyond the end of the century, resisting the forces ranged against it.

London maximized the degree of speciality of land uses. The process had begun well before the nineteenth century but critical anchors were set down. The Royal Exchange, rebuilt between 1841 and 1844 and the Bank of England built between 1788 and 1808 confirmed the specialized financial core of the 'City'. Trading in commodities was filtered out eastward towards dockland where warehousing and industry had been extensively developed after 1800 with the construction of the West India Dock. Between the mixed margins of the financial core, about St Paul's and the legal area which was of long standing, a further specialized area emerged taking its name from an east–west street which crossed a small tributary of the Thames, the Fleet. The nineteenth century saw the growth of a literate public, which, together with the increasing ease of national distribution, rapidly increased the printing and publishing business. Legal publishing, an even more specialized activity, was already located at the intersection of the areas. As Kellett writes, 'Mercery . . . was virtually chased into the West End by the expanding printing business; first the professional publishers of legal and other documents, then the newspapers of Fleet Street' (Kellett, 1969, 301).

In the progressive westward expansion the leader was retailing, and especially the new department store. Shaw has identified three phases in the process (Shaw and Wild 1979, 286). The first was the construction of bazaars in the 1830s and 1840s. There were ten in early Victorian London with Oxford Street, a new area of extension, one of the leading locations. The second was the increase in shop size by amalgamation of properties to give 'island' sites dominating whole blocks. The third phase after 1880 was the construction of department stores, Selfridge's, opened in 1909, has already been noted (p. 93). It symbolized the new significance of Oxford Street. As the retail area extended so, too, were other land-uses dragged westward, both in sympathy with shopping and with the high-quality residential areas. Foremost amongst them were amusement and entertainment which became focused on Soho Square and included the area to the south, where London's theatreland came into being. Again, to the north the section around the British Museum and the University of London developed a character partly derived from

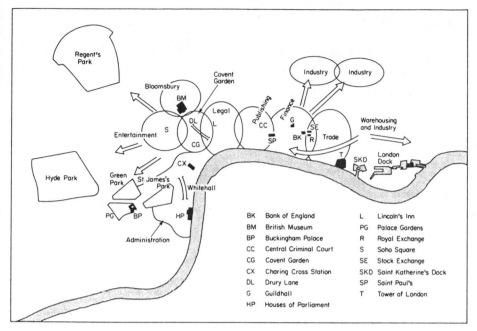

Figure 7.10A London: A generalized diagram of the development of specialized areas. These are shown in the most general fashion and there was a very great amount of overlap of all the uses.

its intellectual environs, partly from the nature of the area to the south and west, and best summarized by the name of one square, Bloomsbury. National administration pushed north from Westminster along Whitehall to meet the western extension at Charing Cross, thus establishing an extensive governmental and administrative area between the river and the royal parks. These specialized areas are indicated on Figure 7.10, but manifestly in none of them was any use exclusive and there was considerable mixing within them and especially in the interstitial areas.

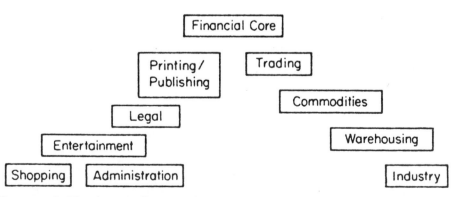

Figure 7.10B This diagram illustrates the way in which more specialized areas progressively broke away from the commercial and financial core, which itself developed 'internal' specialisms.

Conclusion

This chapter has attempted to review the changes which occurred in each of the uses which characterized the central parts of towns and cities and also the spatial relations between them as a relatively undifferentiated core became a complex central business district made up of a mosaic of these uses. It was in the central area that the most characteristic features of the Victorian townscape were to be found, the great warehouses, the magnificent town halls, the department stores occupying whole land blocks. There was, too, a peak of variety and distinction when shops were still individual and distinctive before the common frontages and the chain stores brought a blandness which is now common to most towns. This is how a correspondent of the *Morning Chronicle* described the centre of Merthyr Tydfil in 1850,

> The shops of Merthyr are numerous, well furnished and show all the bustle and activity of a thriving trade. The market-house, which is very capacious, may be termed a 'bazaar of shops'. The scene from six to ten o'clock of a Saturday evening is one of the most extraordinary I have ever witnessed. In this interval what one might suppose the entire labouring population of Merthyr passes through its crowded halls . . . It is not only the field of supply, but evidently the promenade of the working classes . . . one division of the market is appropriated to butcher's meat; another to vegetables; a third to poultry and butter; a fourth to dried stores of bacon, cheese and herrings; a fifth to apples, eggs and fruit . . . There are also stalls of every description for hardware and other shop goods. Hatters, drapers, shoemakers, tinmen, ironmongers and even booksellers, here drive an active and thriving trade . . . Outside the market-house are booths and shows, with they yellow flaming lamps, flaunting pictures, and obstreperous music' (*Morning Chronicle*, 1850).

Perhaps only the innumerable beer houses need to be added to the picture of a shopping centre at mid century when market-hall was still dominant and when the 'promenade of the working classes' already indicated the new nature of the town centre.

8

The demands for urban space: public utilities and institutions

The first question which must arise under the title of this chapter is why there should be a separate consideration of a group of uses only loosely covered by the term public utilities and institutions. It could be contended that they should properly come under the administrative category included within the last chapter. It must be admitted at the outset that definition is hardly precise for within the group are included true public utilities such as waterworks and cemeteries, others which are better called institutions, such as hospitals and asylums, and a still broader collection of uses, such as public parks and gardens and leisure facilities, including golf courses and football fields. It is a motley array of uses with little in common in the way of locational demand. Some, like golf courses, need extensive arcas of land whilst others, like a specialist hospital, will have more intensive needs. Some of the uses can be classed as socially undesirable, like sewage works and gas works or, on a different basis, hospitals for the mentally ill or for patients with communicable diseases. Other uses, such as the golf courses, are socially attractive although they only became popular in England and Wales in the last decade of the nineteenth century.

It is difficult, therefore, to identify a common element in locational terms although one might possible exist in the inability of any of the uses to bid for urban land. The activities of philanthropists or urban managers might affect that factor, but even so, it is a starting point and an important one, for much of the discussion which follows derives from it.

A further common element of association is induced by the relation of the locational characteristics of this group to the way in which towns grew in the nineteenth century. Chapter 3 dealt with urban populations and it was manifest that towns do not simply grow by standard decadal increments. That is equally true of their physical extensions. Towns do not grow outwards by equal extents over successive periods, rather they increase unevenly with periods of rapid enlargement broken by periods of stillstand. During the time when towns do not expand the margin or frontier is often marked by a distinctive feature. That feature will vary in character; it may be a built element such as a town wall; it might be a topographical feature such as a river or steep slope; it might be a cadastral feature such as a land-owner's unwillingness to make land available, especially as prices fall when urban growth is limited. This frontier has been given the name fixation line. Now during a period of relative quiescence those uses which are the subject of this chapter, and generated within the town, will be able to bid successfully for the open land beyond the fixation line and help to create a distinctive fringe belt. The fringe belt's distinction will rest on two aspects. The first is that of land uses already indicated. The second will be the very different pattern of land division, for inside the fixation line the detailed plot divisions of residential and other associated uses will contrast with the much larger subdivision outside the fixation line.

If what has been outlined above corresponds to reality then a distinctive structure will be generated and it is for that reason that a separate consideration of public utilities and institutions in the nineteenth century is justifiable. It is true that after a period of stillstand, and with the renewal of growth and the breaching of the fixation line, the fringe-belt land will be taken into more standard urban uses, the taking over of playing fields for residential development is a well known if always controversial process. But that absorbtion will not eliminate the broad lineaments of the previous land holding pattern, so that fringe belts remain as distinctive structuring elements of towns.

Fixation lines and fringe belts

The concepts of fixation lines and fringe belts were introduced into British urban geography by M. R. G. Conzen (Conzen, 1960) and it has been Jeremy Whitehand (Whitehand, 1967, 1972, 1987) who has broadened their significance by linking them with theoretical interpretations of land uses. A priori, there must be an association between zonal land-use patterns derived both from Burgess's essentially historical schema of town growth and from bid-rent bases. Whitehand's basic contention is simple but it rests on a profound notion which has never had currency in bid-rent interpretations of urban land-use patterns. It is that the bids which users make will not remain static in relation one with the other over time. The standard bid-rent against distance diagram (Carter, 1981, 189–92) represents one condition only when urban growth is rapid. In periods of stillstand when growth is halted, then the bids of developers concerned with land for residential extension will fall. Eventually they will fall below those of other users who want marginal land beyond the fixation line, especially the groups set out at the beginning of this chapter. In this way the fringe belt becomes characterized by a varied but distinctive group of land uses which might well include the large houses in extensive grounds to which the situation outlined also presents an opportunity.

At this point, and before introducing some examples of fringe belts, it is appropriate to consider the nature of institutional uses at some greater length. There has only been one extended study of such uses, that by R. F. Broaderwick (1981) *An investigation into the Location of Institutional Land Uses in Birmingham.* Six categories of institutional uses were examined – parks and open spaces, health institutions, higher education, places of burial, golf courses and a miscellaneous group which includes a wide range of uses such as recreational establishments (sports stadia and racecourses for example), sewage works, exhibition halls, philanthropic orphanages and military barracks. Table 8.1 shows the ratios between these institutional sites and house building during the late nineteenth century. It has been extended into the present century for clarity and continuity. Broaderwick writes in comment on this table, 'the relative importance of institutional sites in the land-use composition should increase during slumps and the four ratios calculated show almost complete agreement with the expected pattern' (Broaderwick, 1981, 214). He adds, however, that the 1891–1901 boom does appear to be an anomaly since it was a period of great institutional activity encouraged by the political and social attitudes of the period. Although he is forced to make many reservations, largely based on the contrast between the intensive–

Table 8.1 Ratios between institutional sites and housebuilding in Birmingham, 1860 –1920 (after Broaderwick, 1981)

		1	2	3	4
1862–1877	Boom	0.85	0.62	0.26	0.59
1878–1890	Slump	1.73	1.24	0.62	1.11
1891–1901	Boom	1.87	0.94	1.13	0.60
1911–1920	Slump	5.74	2.14	4.27	1.47

The ratios are
1 Institutional sites per 1,000 houses built
2 As 1 but excluding parks and open spaces
3 Extensive institutional sites per 1,000 houses built
4 Intensive institutional sites per 10,000 houses built

extensive users of land, and the actions of philanthropists and corporation land managers, a general confirmation of theoretical assumptions is derived.

This is further related to what the author calls the development of an open fringe belt from the 1870s. 'In the southwest quarter of Birmingham, this open fringe belt is almost continuous between South Moseley and Winson Green and included such institutions as the University of Birmingham, The Queen Elizabeth Hospital, Selly Oak colleges, several parks and golf courses, two workhouses and the Borough prison. This open fringe belt is similar in character to those Victorian and Edwardian fringe belts identified by Conzen, Whitehand and Barke' (Broaderwick, 1981, 212).

Against this background Broaderwick proceeds to establish a schematization or a descriptive model of the development of institutional land uses in cities (Figure 8.1). Only the phases relevant to the nineteenth century are included here. Phase One represents the earliest development when institutions, necessarily bound by primitive communications were closely associated with a compact urban core. New developments were possibly a pleasure garden, a bowling green, a burial ground and a workhouse, perhaps with an infirmary attached. In the second phase intra-urban communications were still restrictive but the nineteenth century brought forth a widening range of institutions which took up peripheral locations. Amongst them were botanical gardens, commercial parks, specialist hospitals (the 1845 Lunatic Asylum Act compelled boroughs to provide asylums), municipal cemeteries, religious colleges and adult education centres, often the precursors of universities. In this phase the edge of the built-up area was colonized, especially by those uses needing extensive land areas, or undesirable in central locations. The third phase is dated to the end of the nineteenth century by which date intra-urban transportation had greatly improved. A new fringe of institutions, especially with public parks and hospitals, emerged. Recreational features, such as golf courses, increased greatly in number in England and Wales in the 1890s. In addition, as the fringe extended it merged with those of neighbouring local authorities where similar developments were taking place. At the end of this phase many inner-city located institutions, although firmly established, were facing the problem of removal to the fringe if any additional land area was needed, although that, along with the takeover of large houses for institutional use, is much more a feature of the twentieth century.

Already, however, a disconcerting element has emerged. Between 1770 and 1910 Broaderwick identifies some five slump/boom episodes but if one takes the first phase as representing the position at 1770 then only two distinctive locational

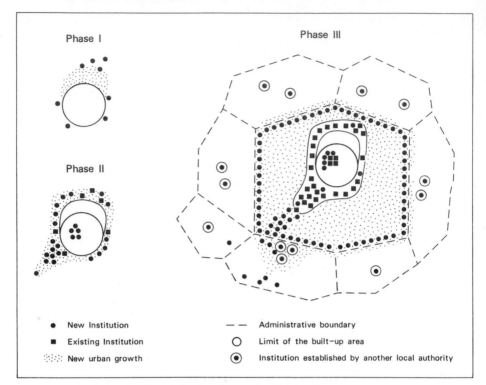

Figure 8.1 A schematization of the development of institutions.
(after Broaderwick, 1981, Figure 7.1)

phases of development are established, a discrepancy which certainly calls into question the acceptance of the basic theory of the relationship between periods of economic recession, institutional uses and fringe-belt development. In order to examine the characteristics of fringe belts in the nineteenth century some examples can be considered.

Whitehand's interpretation of the fringe belts of Newcastle-upon-Tyne are shown in Figure 8.2. In the late nineteenth century Whitehand notes a very clear fixation line to the east of the city in the gorge and valley of Jesmond Dean. The attraction of large residences to that location led to an internal differentiation of the belt with warehouses and industry developing at the Tyne confluence. Also during the Victorian period he adds into the picture the construction of various institutions, notably several isolation hospitals and a lunatic asylum, as well as cemeteries and waterworks. 'Where the geology permitted quarries and brickworks found similar peripheral locations. During the vigorous residential growth of the late Victorian period some of the less distant plots were swallowed up by the house builders, but the majority survived to form a discontinuous belt of varying width. This belt stretched northward from the riverside in the area of Elswick and Benwell and incorporated the open spaces of the Town Moor and Nun's Moor . . . In the east the belt followed approximately the line of Jesmond Dene' (Whitehand, 1967, 224) where the distinct fixation line has already been noted.

It is useful to add two points relating to common nineteenth-century fringe-belt uses. Town Moor, of course, was part of the town's common land and these open spaces were still available at the end of the nineteenth century when soccer was becoming organized into a national league. Characteristically, therefore, modern soccer stadia occupy fringe-belt land. Turf Moor at Burnley is a clear parallel. Again, at Cardiff, the Arms Park and the national stadium, although on reclaimed river land, did form part of the inner fringe belt, as did the land already discussed, on which Cathays Park was developed. Because of its very late development the inner fringe belt at Cardiff is a nineteenth-century feature. The second characteristic use is derived from the fact that many universities were founded during the late nineteenth century and they, too, took over available land. Reference back to Birmingham will show that Mason College, as the university then was, occupied what Broaderwick described as 'the open fringe belt'.

The second example is the much smaller town of Aberystwyth in West Wales. The physical structure of Aberystwyth during the nineteenth century can be related to the existence of two fixation lines and two fringe belts (Figure 8.3). The inner and first fixation line was that of the medieval town walls, together with the physical limits of the small, extra-mural bridgehead settlement of Trefechan (literally 'little

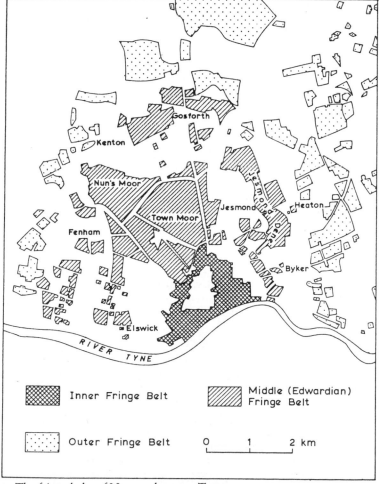

Figure 8.2 The fringe belts of Newcastle-upon-Tyne.
(after Whitehand, 1987, 80, Figure 5.2)

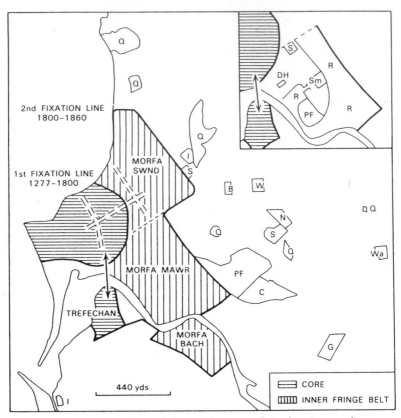

Figure 8.3 Fixation lines and fringe belts in Aberystwyth in the nineteenth century. Inset, Morfa Mawr, fringe uses at the end of the century. The streets of the CBD in 1900 are indicated.

I Infirmary; S School; W Union Workhouse; Wa Waterworks; G Gasworks; C Cemetery; PF Playing Fields; N Nursery; Q Quarry; R Railways; Sm Smithfield; DH Drillhall.

town'). Apart from Trefechan, no houses were built outside the walls until late in the 1790s, although minor encroachments had taken place during the eighteenth century. The medieval, or inner first fixation line lasted therefore for some 500 years. To a large extent it remained inviolate because the town itself was but a skeleton of streets, and a good deal of open land remained within the walls, even in the second half of the nineteenth century. Beyond this fixation line lay the Inner Fringe Belt. It was composed of the extra-mural common lands of the borough, which were made up of three tracts of marshland which surrounded the small hill on which the town had been established in 1277. These were Morfa Swnd (Sandmarsh), Morfa Mawr (Great Marsh) and Morfa Bach (Little Marsh). Since there was ample intra-mural space, little specifically urban use was made of these lands, other than the location of the cattle market, the pound and the town gallows! Fringe uses, therefore, did not characterize them in any distinctive way, other than by the manner in which they contributed to those agricultural activities in which the burgesses themselves were involved. For small, remote towns in a period before effective transport, the provision of food can be regarded as a distinctive fringe use, although the conventional view of such uses is one which is dominated by the growth of nineteenth-century urban institutions. It was not until 1813 that these lands of the Inner Fringe Belt were formally divided and leased. In that year an entry in the Court Leet recorded, 'We, the jury, direct that part of the waste land called Morfa Swnd be mapped and divided into convenient spots for building'. The trigger for this decision was the demand for land brought about by the first growth phase as a resort town at the turn of the eighteenth century. Morfa Swnd was enclosed, divided and leased, as was Morfa Mawr, but the latter remained in agricultural use.

With the extension of the town in this manner, a new and second fixation line was created. This was at the limit of borough common land, which had previously been classed as marshland, and which was clearly marked by the steep slopes of the Rheidol Valley sides both to north and south. Beyond this line, land was in private hands, mainly those of two prestigious local families, the Pryses of Gogerddan and the Powells of Nanteos. It is also interesting to observe that at two points a use, derived from the seaport role and demanding extensive linear land areas, marked the new fixation line, for ropewalks had been established both to the north of Morfa Swnd, where a lease had been granted in 1778 and to the south of Trefechan (1810). Land ownership, land tenure, a specific land use and sharp breaks of slope all contributed to the emergence of this new fixation line, which was to last from the early nineteenth century, when the development of the Inner Fringe Belt first began by encroachment, until the 1870s when, after the coming of the railway in 1864, the second phase of population growth and physical extension pushed settlement beyond it and into the Middle Fringe Belt. This Middle Belt, beyond the second fixation line, was formed during the middle and later part of the nineteenth century when, related to a whole range of social legislation, towns were generating a variety of associated institutions. It had, therefore, to a much greater degree, the characteristic uses of a fringe belt. It was dominated by a series of quarries, worked into the valley side and developed as a source of stone for the phase of building after 1813. It also included the Union Workhouse and a militia barracks from the middle of the century and, by the end an infirmary, gas works and the town cemetery, as well as schools and playing fields, including, characteristically, the town soccer ground (Figure 8.3). After 1880 these were interspersed with the extending frontier of house building, as sites on the northern valley side were used. It is difficult to place Morfa Mawr in this context. Physically it was part of the Inner Fringe Belt, but it remained undeveloped until the later part of the century when it was characterized by Middle Belt uses, including recreation grounds, railway yards, the smithfield, a drill hall and a school. To the south of the river the situation was simpler, for the very steep Rheidol slope meant that the first and second fixation lines were coterminous; there was no distinction between an Inner and a Middle Fringe Belt.

By the end of the nineteenth century, therefore, the town was structured into three distinctive areas – Core, Inner Fringe Belt and Middle Fringe Belt. The physical evidence of the first fixation line, the town walls, and of the second, the ropewalks, had been removed, but the influence of the two lines was clearly apparent. The physical structuring of the composing elements remains quite clear. The intricate block make-up of the core, the grid-like streets of the Inner Fringe Belt and the linear extensions through a much more open patterning of the Middle Belt are easily recognizable on a present-day map. It is perhaps worth adding that at Aberystwyth the university began not in the Inner Fringe Belt where it might have been expected but in the core itself by the takeover of an existing, or at least a half-finished, building. But the constriction of space led to the College's purchase or lease of buildings in the Inner Fringe Belt until it could leapfrog out into pre-erved open land of the Middle Fringe Belt where it constitutes a characteristic institutional use. The National Library of Wales followed a similar path.

It is apparent from the above examples from towns of greatly varying size that the fringe belt was a very distinctive feature of the nineteenth-century town. It was a product of necessity brought about by the quite massive increase of institutional use of all sorts during the century, and of opportunity, as land prices fell during periods of recession, enabling lower-order bidders for extensive areas to enter the land allocation process.

Problems of the fringe-belt concept

Although a convincing case has been made for the significance of fringe belts, derived from the special locational needs of institutional uses, it is not one which has gone without challenge.

The first problem is that not all institutional uses have the same locational demands. Whereas extensive users, such as golf courses, look clearly to the fringe, for intensive users accessibility is a significant desideratum. Thus specialist hospitals, for example, will seek sites in the inner city. It is not possible simply to equate all institutional users with fringe-belt locations. It can be contended, therefore, that the location of institutions and public utilities and the process of fringe-belt formation are two associated but separate problems.

The second point of criticism has already been introduced (page 114). Given the alternation of boom and slump during the nineteenth century there should be a whole series of correlated fringe belts rather than the conventional two, usually called Inner and Middle, which are seen as nineteenth-century features. In short, there is seemingly no agreement in the wave-length of boom–slump occurrences and that of the creation of fringe belts. This is a criticism of the explanation rather than of the reality. Certainly local factors were paramount and operated at a much cruder scale than national economic cycles. At Aberystwyth there were two significant growth phases, one relating to the initial rise of the resort function at the beginning of the nineteenth century, the other to the arrival of the railway and the reinvigoration of the same role in the 1860s. It was these larger-scale, local happenings which were significant, not the national sequence of upturns and downturns, although they were necessarily linked.

A third criticism again relates to the theory of explanation. The operation of the bid-rent explanation must depend on an absolutely free market. But the market never was free. In the private sector landowners could distort the selling and buying of land, whilst in the public sector urban managers could do the same. But that is applicable to all land-use models and was in reality a distortion rather than an overturning.

A direct refutation of the basic principle, and hence implicitly of the location of institutions, has been made by Homan and Rowley (1979). Their basis is similar to that of the third criticism offered in the paragraph above. 'It is suggested that direct attempts to reconcile urban economic-rent theory and the growth of nineteenth-century British cities are misplaced, as non-economic considerations play a fundamental role in the decision-making processes of land developers.' (Homan and Rowley, 1979, 137). Their investigation considers the building of churches and chapels in Sheffield and leads to the conclusion that it was part of the process of residential extension rather than something separate and related to separate locations. It is a tenuous basis on which to refute the whole of Whitehand's work, especially as parallels with educational institutions and parks are suggested but not pursued. Moreover, there is a real dichotomy between churches with and without burial grounds for the latter group required greater land area and unless a distinction is made, as it is by Broaderwick in his work, then the value of the conclusions is limited. Even so, a valid point is made by Homan and Rowley. It is possible to propose the complete reverse of the theory behind fringe-belt formation, that institutions proliferate when residential growth is booming and come into being when a necessary threshold of demand has been reached. Broaderwick had to note an exception in his analysis for that very reason (page 112).

There is a final criticism of the fringe-belt concept but one not so relevant to the nineteenth century, that is, that uses are eventually changed and the land holdings modified, so that the belt is absorbed and loses its identity. This is denied by both Conzen and Whitehand for they see the characteristics of the fringe belt indelibly written into the town's morphology. Certainly, as has been suggested, most fringe belts can still be identified at the present.

Conclusion

This chapter has been written under the heading of institutional uses and public utilities. The fringe-belt concept has been introduced as a means of interpreting the location of such uses. It is a useful but only a partial basis. Institutions and public utilities were not homogeneous, the starting point for Homan and Rowley, and a necessary premise for Broaderwick. Homan and Rowley emphasize the contrast between neighbourhood and city-wide service, but also critical was the amount of land needed, a factor absent from the simple bid-rent model. The very nature of nineteenth-century social development engendered a mass of institutions and utilities, it was one of the characteristic features of the age. Some of them sought inner-city locations where city-wide service was most successfully offered. Others were neighbourhood related, like churches and schools, and were constructed as part of residential developments. But there was also a distinctive group which, by the nature and/or amount of land demanded, looked to fringe locations and created those fringe belts which are a common and distinctive feature of the nineteenth-century city.

9

The demands on urban space: residential areas

Urban population growth has already been examined in Chapter 3 in aggregate terms, at the level of the urban system, and that highlighted the dramatic changes which occurred in Britain as it was transformed from a dominantly rural to a largely urban nation. This chapter is also concerned with urban population, but this time at the intra-urban level. It discusses the residential patterns which emerged as people with different occupations and backgrounds jostled for space within expanding towns.

The nineteenth century can be seen as very much a century of change in intra-urban residential patterning. In the simplest terms, towns gradually lost the vestiges of their pre-industrial structures and began to take on the internal characteristics of the modern industrial city. This meant the abandonment of housing in central or near central locations by the more affluent in society and the colonization by the middle classes of the newly emerging suburban fringe. It also meant the appearance of large-scale segregated residential areas according to socio-economic, family or ethnic status, in contrast to the much finer or small-scale patterns of segregation in pre-Victorian towns where the different status dimensions were often interrelated. Again, at its simplest, this change has been seen as the demise of the structure associated with Sjoberg's pre-industrial town and the rise of Burgess's ecological city (Lewis, 1979).

The legacy of the past which survived into the nineteenth century has been succinctly described as follows:

> elements of the spatial structure proposed by Sjoberg and Vance still existed at the beginning of the nineteenth century. At least some of the elite – merchants, solicitors and surgeons, for example – continued to live in fashionable town-houses in the centres of cities, often combining their residence with their place of work. For most people segregation existed only at the scale of the individual dwelling; in fact, vertical segregation within each dwelling, whereby social status decreased as you ascended from the master's rooms on the first floor to the servants in the attic, or descended from the ground-floor shop to the apprentices in the basement, was often more important than horizontal segregation between adjacent houses or streets. It was rare to observe whole districts of uniform social status, except for the segregation of the elite at one end of the spectrum, and immigrant groups – usually Irish or Jewish – at the other. (Dennis and Clout, 1980, 62)

Dennis and Clout presented some of these elements in a simple model, to which they added the harbingers of change, namely the middle-class suburb and industry (Figure 9.1A). Over the course of the century this pattern was transformed into the

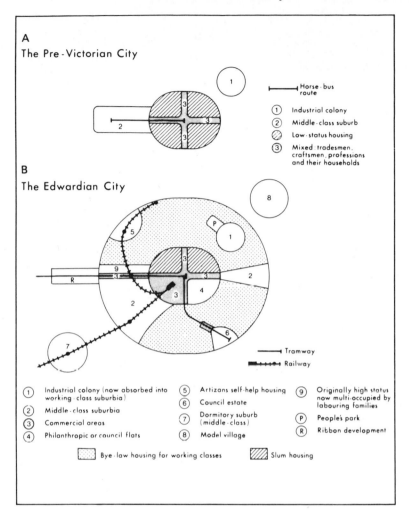

A
The Pre-Victorian City

	Horse-bus route
(1)	Industrial colony
(2)	Middle-class suburb
(☉)	Low-status housing
(3)	Mixed: tradesmen, craftsmen, professions and their households

B
The Edwardian City

| | | Tramway |
| | | Railway |

(1)	Industrial colony (now absorbed into working-class suburbia)	(5)	Artizans self-help housing	(9)	Originally high status now multi-occupied by labouring families
(2)	Middle-class suburbia	(6)	Council estate	(P)	People's park
(3)	Commercial areas	(7)	Dormitory suburb (middle-class)	(R)	Ribbon development
(4)	Philanthropic or council flats	(8)	Model village		

Bye-law housing for working classes   Slum housing

Figure 9.1 A model of evolving urban structures: **A** the pre-Victorian city; **B** the Edwardian city.
(after Dennis and Clout, 1980, Figure 5.7, 83)

segregated areas shown in Figure 9.1B. While small-scale segregation existed in the pre-industrial city, spatially between front streets and back streets and vertically between the floors in households, during the nineteenth century larger-scale segregated regions appeared. Behind this transformation lay the processes which were shaping Britain's industrial society, including

1   industrial change, involving the building of large factories and other works set in extensive areas of land, and housing for the workers (Chapter 6);
2   changes in retailing and the professions, by which the mixed residential and commercial cores of towns progressively became specialized central business districts with lock-up premises (Chapter 7);
3   improvements in transport, which facilitated the outward spread of cities;

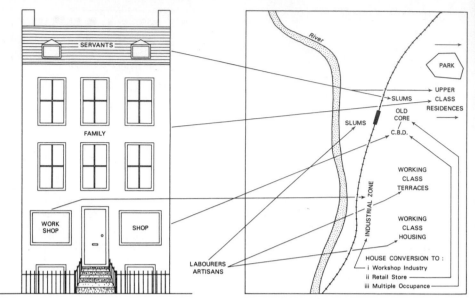

Figure 9.2 A model of urban spatial transformation in the nineteenth century. (after Carter, 1980, Figure 2, 183)

4   improvements in household amenities and the growing favour of the suburban villa over the Regency town house;
5   the evolving class structure; and
6   the mixing of migrants from different backgrounds and perhaps cultures.

From the geographical point of view, the outcome of these changes

> can be summarized, however inadequately, in the two words 'separation' and 'segregation'. Making and selling (or manufacturing and retailing) were separated, and the consequence was a segregation of the functions, in spatial terms, into industrial areas and shopping centres. Working and living became separated and the journey to work initiated. But those who could pay the cost could most effectively separate themselves from the pollution of industry and the increasing noise, bustle and dirt of the shopping centre. Those who could not afford this separation were constrained into living in those parts adjacent to the industrial areas. Segregation, based on social class or ability to pay, thus became visible because it appeared at a clearly discernible scale. The rich moved to the new marginal suburbs where houses built to the demands of new technology and new taste could be erected. (Carter, 1980, 182–4)

This has been presented in an elementary model (Figure 9.2), in which all of the components and personnel of the large household in the pre-industrial town (the workshop and shop, the employees, the employer and his family, and the servants) are separated out and are reconstituted in the industrial city, not under one roof but in their own distinctive urban regions. In effect, what Figure 9.2 seeks to do is to give the dynamic link between A and B in Figure 9.1. While these diagrams provide a useful shorthand description of what was taking place, it has to be recognized that the processes of residential sorting and sifting which occurred as individuals and families found their niches in urban society were much more complex. However, it is possible to make headway in understanding their operation by looking at some of the key factors at work, in particular, the evolving class structure, ethnicity, and mobility.

Class

One of the major determinants of residential segregation was class. Class is a nebulous and emotive word which has generated a considerable and continuing discourse on all its ramifications, but for the purposes of discussion here, it can be interpreted as the 'status' of persons or households in society, derived partly from their *income* and partly from the *esteem* in which their occupations were held by their contemporaries. Class was translated into geographical patterns through the clustering of those people from the same stratum in society into their own localities. Thus, at the most basic level, differentiation by class could lead to the formation of 'upper', 'middle' and 'lower' class districts. However, one of the consequences of industrialization was not such a simplification of the divisions in society, but rather their elaboration and extension through a burgeoning in the variety of occupations. Not only was there a proliferation of specialized activities in the manufacturing sector, but also in transport, commerce, the professions and administration. These occupations displayed increasingly complex patterns of differentiation within which the various groups or classes sought to set themselves apart from each other through their distinctive life-styles, social activities and places of residence. In short,

> an urbanized society results in a competitive system generating a complex stratification deriving partly from the economic rewards of skilled tasks, both manual and pro-fessional, and partly from the esteem in which those tasks are held. Those having a status defend it, particularly against dilution. That defence was apparent in elaborate initiation rituals and became part of the Trade Union movement's central concern. Those lacking status sought to achieve it and, since the basis was competition, upward mobility was characteristic: the term *nouveau riche* expresses the reality of such upward mobility. Each stratum of society sought to distance itself from that below and to copy the life style of that above. (Carter and Lewis, 1983, 57–8)

At the top of the social hierarchy were the upper classes, traditionally drawn from the landed nobility and the gentry but now expanded by those who had become very wealthy through urban trade. This elite group was the subject of Davidoff's book '*The best circles: society, etiquette and the season*' (Davidoff, 1973). Their importance in urban society lay not in their number, which was relatively small, but in their life-style, attitudes and behaviour which set standards to be copied by those lower in the social hierarchy. Life was lived through an elaborate code of etiquette which effectively maintained the exclusiveness of the elite. Spatially, too, they were able to put up barriers around their communities, as shown by Davidoff's observation on West London during the 'season'.

> The building of this sense of community started early in childhood. For example, Hamilton Gardens, which was a section of Hyde Park enclosed by iron railings and a locked gate to which only select residents of Mayfair had keys, was used as a private playground for upper-class children. During the week in the Season, Nannies over-looked the children of the aristocracy playing together with the offspring of those professional men, bankers and other gentlemen whose wealth and family ties gave them admittance to these circles. This pattern of private communal gardens was imitated on a lesser scale all over the West End. (Davidoff, 1973, 31)

Below this upper stratum which was distinguished by nobility and wealth, there was a gradation down through the lower orders of society. While it is convenient to just use the two broad divisions of 'middle' and 'lower' classes, in reality there were much finer demarcations, perhaps determined as much or more by perception as by wealth. Within the artisan working class, for example, Crossick (1978) has shown

Table 9.1 Weekly earnings in specific occupations, Deptford, London 1887 (after Crossick, 1978, Table 6.1, 110)

Occupation	Cases	% earning over 30s.	% earning under 21s.
Shipwrights and boatbuilders	102	81.4	7.9
Skilled engineering and metal	248	80.2	8.8
Building crafts*	242	77.7	7.1
Carpenters, joiners, sawyers	368	70.4	13.2
Furniture makers	38	68.4	23.8
Commercial clerks	151	64.2	17.9
Printers	67	62.7	12.0
Smiths and tinworkers	526	62.4	14.4
Painters, plumbers	278	56.8	15.6
Tailors	32	40.6	40.6
Dock labourers	290	26.9	40.0
Bakers	46	23.9	28.3
Shoemakers	82	14.6	66.9
Labourers	2,257	7.7	54.1
Carmen, carters	274	6.9	41.6

Notes: * excluding carpenters, joiners, painters.
N.B. 16.5 per cent of those interviewed refused to answer this question. The question asked was, 'If in work at present, weekly wages or earnings. If out of work at present, ordinary weekly wages when in work.'
Calculated from Conditions of the Working Classes, *Parliamentary Papers*, 1887, LXXI.

that a pecking order emerged, with the highly skilled workers standing out as the labour aristocracy, at the head of those employed in manual work. Table 9.1 sets out the weekly earnings of specific occupations in Deptford in 1887, and shows the wide gulf that existed between the skilled and semi-skilled and those who were general labourers. As Crossick has argued, the bases of working-class stratification were derived from the workplace.

> The general exclusiveness prevailed, particularly between the better-skilled men and the unskilled. 'There is no place', wrote an anonymous 'Working-man' of his fellow London workers (in 1879), 'in which class distinctions are more sharply defined, or strongly, or if need be, violently maintained than in the workshop . . . Evil would certainly befall' any labourer who tried to assume equality with an artisan. (Crossick, 1978, 129)

This was translated into physical separation through housing, with the artisan elite expressing their independence and respectability through their own residential districts. 'In Deptford they moved to the New Town and Hatcham, in Greenwich to streets in the Maze Hill and Greenwich Park areas, in Woolwich to Plumstead New Town and Burrage Town . . . Moving to status-defined residential areas intensified social attitudes, for it added a spatial dimension to social separation' (Crossick, 1978, 145). As has been pointed out elsewhere, 'these new houses were not necessarily well built, nor did moving to them necessarily mean leaving poor housing conditions. What they did offer was respectability symbolized in architecture and socially acceptable neighbours' (Carter and Lewis, 1983, 59). Thus class, through the members of the various strata seeking to distance themselves from those below and to acquire some of the refinements of those above, was one of the determinants of residential segregation.

As indicated in the consideration of data sources in Chapter 2, it has been commonplace for geographers (and others) to use occupation as the main measure of class or status. Although there are limitations in such a shorthand approach, it has the advantages of giving an indication of income (and thus the life-style which could be afforded) and of being readily available in the census enumerators' returns. Ideally, as Dennis has argued (1984, 196), it is most desirable to use occupation in association with information on attitudes and behaviour, but, in general, such information is piecemeal for individual towns whereas occupation can be readily used as the one standard measure of differentiation, both within towns and for comparisons between towns. Certainly, the great majority of geographical studies of residential differentiation by class have relied on occupational data to delineate the broad patterns.

The geographical literature contains numerous empirical studies which demonstrate that British towns displayed segregated residential districts, though there is debate over the timing of their appearance. One of the most comprehensive surveys

Table 9.2 Liverpool 1871: loadings on the first three components (after Lawton and Pooley, 1976, Table 8)

	Component 1		Component 2		Component 3	
Percentage of variation accounted for	24.6		13.6		9.3	
Cumulative percentage	24.6		38.2		47.5	
Loadings of variables						
Positive	Population density	.78	Socioecon. gp. 4 & 5	.95	Age 15–64	.89
	Housing density	.75	Socioecon. gp. 5	.75	Lodgers	.51
	Residential land	.70	Socioecon. gp. 4	.53	Servant index	.38
	Resident in courts	.68	Resident in courts	.43	Servants	.33
	Multiple occupance	.64	Irish	.41	Women in workforce	.33
	Irish	.64	Houseful size	.37	Widowed heads	.31
	Houseful size	.46				
	Socioecon. gp. 4	.43				
	Socioecon. gp. 5	.43				
	Lodgers	.33				
Negative	Distance from centre	.78	Socioecon. gp. 3	.92	Age 0–14	.92
	English migrants	.74	Residential land	.39	Age 0–4	.76
	Socioecon. gp. 1 & 2	.49	Welsh migrants	.37	Nuclear family	.59
	Servants	.49	Scots migrants	.36		
	Servant index	.47				
	Houseful size	.33				

Note: Variables are in percentages except for distance and densities. The servant index is servants/ nuclear and extended family × 100.

of a large city is that by Lawton and Pooley (1976) of Liverpool in 1871. Using a set of 35 variables, largely derived from the census, for 394 sub-areas (made up of revised enumeration districts), they carried out a detailed examination of the distribution of individual variables and a number of multivariate analyses of the total data set. From the latter, a principal component analysis with varimax rotation was selected to describe the spatial structure of the city. The three most important components, accounting for 24.6, 13.6 and 9.3 per cent of the variation, can be broadly described as socio-economic status, ethnicity and life-cycle or family status (see Table 9.2 for the loadings of variables on each of these components). Component One separates those areas of high occupational status, good quality housing, low density occupancy, servant-keeping and English migrant origin from those districts, mainly in the town centre, characterized by poor quality housing, overcrowding, low occupational status and high proportions of Irish migrants. 'This factor indicates the main line of differentiation within mid-Victorian Liverpool, a dichotomy between high-status servant-keeping areas and central areas of decaying overcrowded housing, occupied by the Irish and avoided by migrants from England' (Lawton and Pooley, 1976, 45). It should be noted, however, that this component is a little wider than just socio-economic status or class in that there is a strong ethnic element as well, in particular a positive association with Irish and a negative association with English migrants. The second component isolates a working class and ethnic dimension, loading highly on measures of socio-economic status and migrant groups; it can be labelled as ethnicity. It contrasts the unskilled and, to a lesser extent, the semi-skilled, working population, a large proportion of which was Irish, to the skilled working-class population which was associated with Welsh and Scots in-migrants. The third easily identifiable component is a family status dimension. This distinguishes between families of more mature age structure, in which the economically active age groups were dominant and where there were many co-resident lodgers and servants, and families at an earlier stage in the life cycle, consisting primarily of nuclear families with children. These results led Lawton and Pooley to conclude that 'at a general level at least, the urban structure of mid-Victorian Liverpool was similar to that of the many modern cities which have been studied . . . Family status, socio-economic status and migrant status were each important elements in Liverpool's social geography' (Lawton and Pooley, 1976, 46). What is clear is that class was the major factor of differentiation, with the high-status, servant-keeping families separated from the mass of the middle and working classes, and the working classes subdivided into the skilled and unskilled.

The results of this analysis were used to expose the spatial structure of the city, to determine whether the 394 sub-areas formed distinctive residential districts. This was achieved by looking not just at socio-economic status but at all three components. The linkage procedure employed identified 20 contiguous areas which had broadly similar social characteristics, but which were separated by several zones of transition. The spatial structure revealed, represented schematically in Figure 9.3, is summarized by Lawton and Pooley as follows:

> Though there is a complex mixture of sectors, zones and nuclei, a series of distinct elements emerge. First, the older central residential districts, approximately within the 1851 built-up area, had developed by 1871 into a series of distinctive sectors, largely constrained by the location of the docks, the principal routeways radiating from the centre and surviving islands of high-status property. Secondly, outside the 1851 built-up area broadly zonal characteristics are dominant: population from the central area moved out into the newer suburbs within which there were several areas of a transitional nature. Complementary to the mainly outward direction of internal

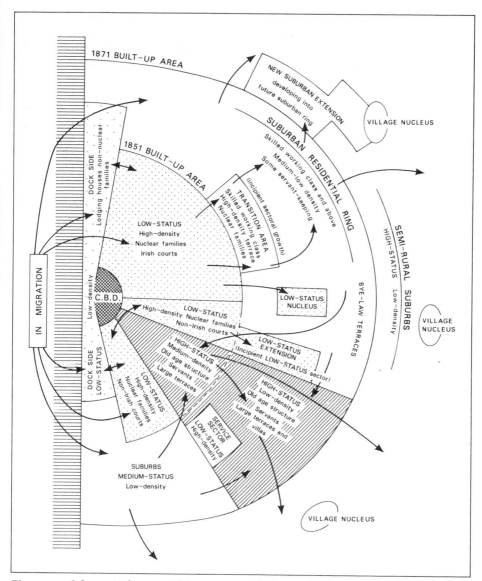

Figure 9.3 Schematic diagram of the structure of Liverpool in 1871.
(after Lawton and Pooley, 1976, Figure 26)

population movement, the central zones were largely fed by in-migrants, some of whom were to stay permanently in low-status areas, others of whom soon moved out to the newer suburbs. There was also considerable movement between sectors of low-status housing, often caused by displacement due to urban redevelopment, though non-Irish families generally avoided the Irish areas. Thirdly, the high-status population moved sectorally into the third peripheral zone whilst, within the suburban ring, lateral movement took socially upwardmoving families into the high-status sector to the south. In addition to these distinctive movements, each area gained by in-migration at various levels of the social stratum. (Lawton and Pooley, 1976, 51)

Thus it can be argued that by the 1870s the city had taken on a modern form, with evidence of segregated residential districts.

In his study of Wolverhampton, Shaw (1979) has also shown that distinctive residential areas had emerged by the 1870s. In this case there was a major contrast between the east and west of the town: the east was dominated by mines and derelict workings, factories and warehouses whereas the west contained much open agricultural land and parks. Working on a framework of 200m grid squares, Shaw plotted the distribution of household heads in classes I and II in 1871 (see Figure 9.4A) and of the more substantial residential properties, those he labelled as types C and D houses in his classification (Figure 9.4B). It is apparent from these

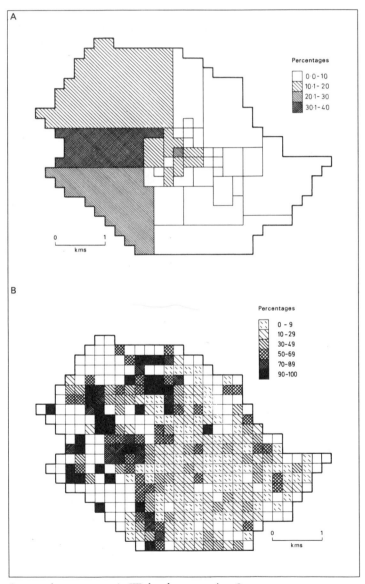

Figure 9.4 Intra-urban patterns in Wolverhampton in 1871.
A Percentage of household heads in classes I and II.
B Proportion of housing types C and D.
(after Shaw, 1979, Figure 4, 198, and Figure 7, 204)

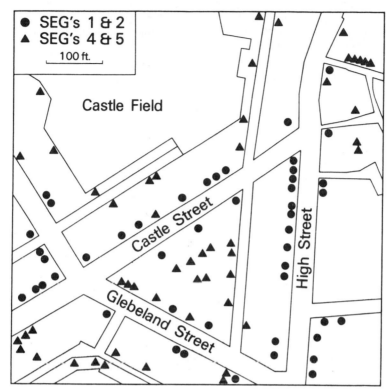

Figure 9.5 Distribution of household heads in highest and lowest socio-economic groups within one central grid square in Merthyr Tydfil in 1851.
(after Carter and Wheatley, 1982, Figure 32, 117)

maps that by this date a pronounced social contrast had emerged between the two halves of the town, with the more rural west attracting the higher social classes and the better residences, and conversely the south and east dominated by poorer social groups and smaller houses. 'To judge from the distribution of those employed in potentially smokey, noisy and unattractive manufacturing trades, the effects of industry were almost certainly greatest in the south and east. It is therefore no surprise that those with any choice opted to live in the expanding western and north-western suburbs' (Shaw, 1979, 206).

It is necessary to introduce a note of caution over the identification of seemingly homogeneous residential areas through the aggregate analysis of census data. Most districts were far from homogeneous in their social composition. The concentration of members of a particular class in a residential area was not to the exclusion of other classes. For example, in Carter and Wheatley's study (1982) of Merthyr Tydfil in 1851 the town centre emerged as an area of high status, but it contained a considerable variety in terms of its class composition as can be seen from Figure 9.5 which is a 200m grid square from the centre of the town showing household heads by socio-economic group. In its detailed composition it had as many household heads in socio-economic groups 4 and 5 as in groups 1 and 2. However, within the centre the two groups did not occupy the same locations, for whereas the high-status household heads were situated in the main streets (High Street and Castle Street), those in the lowest groups were found in the back streets, alleys and courts which lay behind the main thoroughfares. While the trend in the nineteenth century was towards more homogeneous residential areas, as demonstrated by aggregate analyses, in their detail these districts often contained mixing and finer patterns of spatial differentiation.

The observation that the trend was towards class-based residential districts raises the question as to when the trend became apparent in British towns. Did these segregated areas emerge in the early 1800s, sweeping aside any remaining vestiges of the pre-industrial structure, or did they become features of cities in the later decades of the century? This question has led to controversy in the geographical literature, with some writers arguing the former and others citing evidence in favour of the latter. On the one hand, Cannadine (1977), for example, has claimed the early appearance of segregated districts, at least for the well-to-do. He used a number of studies to argue that there was an 'unprecedented degree of residential segregation in mid nineteenth-century England', pointing out that

> the elite of Leeds, after a false start in the Park, settled down to enjoy the suburban delights of Headingly and Chapeltown, areas which, at three to four miles from the town centre, required transport to reach them, thereby putting them effectively beyond the range of all but the wealthy. And what was true of the merchant princes of Leeds also held good for their cousins in Oldham, Glasgow, Nottingham, Manchester and Liverpool who, sheltered behind the barriers of distance, toll gates, park keepers and restrictive covenants, were able to enjoy a life of segregated quietude, surrounded by parks, pleasure grounds, botanical gardens and exclusive schools. (Cannadine, 1977, 464–5)

Cannadine then went on to look at the attempts by Lord Calthorpe and his agent in the early part of the century to develop the Edgbaston estate in west Birmingham as an exclusive suburb for the wealthy.

Of course, this quest for privacy and seclusion was something which had been displayed by the affluent in London when they occupied their new fashionable town houses in the late eighteenth and early nineteenth centuries. Here, at the extreme, exclusive neighbourhoods were sealed off behind barriers. One such example was the Bedford estate in Bloomsbury where lodges and gates were built at the main entrances. 'There were five of these Lodges, each tenanted by a liveried gate-keeper, whose job it was to keep the more unsavoury type of traffic from the Estate. Thus omnibuses, carts, droves of cattle and low persons generally were excluded, but pedestrians, carriages and gentlemen on horseback were allowed through. The Estate authorities would have liked to exclude hackney carriages to Euston Station altogether, but a healthy respect for the railway lobby in the House of Commons prevented them, though they were not allowed through the Upper Woburn Place gate as late as the 1840s' (Hobhouse, 1971, 74). Likewise the nearby Foundling estate felt a similar need to insulate itself from its neighbours to the north and east, and barriers were installed on the Grosvenor estate in Belgravia where bar-keepers with 'glazed hats and gold hatbands' were employed to exclude heavy traffic (Hobhouse, 1971, 127). Despite the inconvenience caused to hackney carriages, hauliers and tradesmen, gates and bars were strongly defended by the protected residents right to the end of the century. Their removal came under Acts of 1890 and 1893 which authorized the new London County Council to open up the thoroughfares. Both in the fashionable squares of the metropolis and in the newly-built estates of villas on the fringe of Britain's largest cities there were segregated residential areas for the wealthy early in the century.

While the most affluent in large cities may have lived in their own districts relatively early in the century, it has been argued by some writers that the more widespread appearance of segregated residential areas for the majority of the population came relatively late. Ward has been a consistent proponent of a very late emergence of such segregated living (Ward, 1975). He has suggested that 'apart from the exclusive precincts of the wealthy and socially prominent, and highly localized concentrations of the poverty-stricken, most new sections of early and mid nineteenth-century industrial cities housed people of diverse occupations which recorded limited but significant differences in remuneration and status' (Ward, 1975, 145), and that 'until quite late in the nineteenth century, "modern" levels and kinds of residential differentiation were quite weakly developed' (Ward, 1975, 151). Even for the most wealthy, there is a great deal of evidence which shows that the better-off residents in many towns continued to live in or near the central core until later in the century. Small industrial towns and those towns which experienced relatively late industrial growth seem to have retained the residence of those engaged in trade and the professions at their hearts, thereby displaying spatial patterns reminiscent of the pre-industrial form. In his examination of Chorley in the middle of the century, Warnes found that the central area was distinguished by relatively high numbers of the upper-status groups and an unusually low density of the unskilled (Warnes, 1973, 183). Likewise Lewis has shown that the ports of Cardiff and Newport – both of which were relatively late developers – displayed traditional residential patterns until the second half of the century, with the main streets and their immediate offshoots housing the most affluent. In Cardiff the T-shaped axes of the old medieval town were picked out by data derived from rate books and the 1851 census, as was the eastward extension into Crockherbtown and Charles Street (Lewis, 1979). This was corroborated in H. J. Paine's Public Health Report for 1855 in which the east–west axis of the old town was seen to be 'inhabited principally by gentry, professional men, and respectable tradesmen' and the north–south axis of St Mary Street was 'inhabited by respectable tradesmen and others' (Lewis, 1979, 138). In Newport, too, Lewis's analysis of rateable values in 1854 identified the main thoroughfares as the locations of high value premises (Lewis, 1985a). There were two particular concentrations. The first was the main commercial streets of the growing town, encompassing High Street, Commercial Street and Commercial Road with offshoots in Llanarth Street and Great Dock Street. By and large, these properties contained both commercial and residential quarters for shopkeepers and professionals who were still resident at their business premises. The second comprised the clusters of high-status town houses on the old medieval axis of Stow Hill, or near it, in Victoria Place, Clifton Place, York Place and Cambria Place. A fine example was Victoria Place, comprising two short terraces of six Georgian style houses (Plate 9.1). The superiority of these houses was confirmed in contemporary accounts. Clark's Report to the General Board of Health in 1850 described the houses on Stow Hill as 'of a better sort', of Commercial Road it was observed that 'like most of the principal streets, this is broad, and contains many excellent houses', and in Great Dock Street there were 'nineteen houses of a superior class'. A later observer wrote about this last area that 'for quite half a century this locality was considered to be the most fashionable place of residence in Newport'. Attention was also drawn to the superiority of the short, upper-class terraces near Stow Hill: 'The erection of Victoria Place – an elegant street, branching from Stow-hill to High-Street – took place soon after the completion of the Dock . . . The spirit of improvement thus awakened, resulted in the building of Clifton Place, a commanding terrace adjacent to the Cemetery' (Lewis, 1985a, 138–41). Thus, these examples demonstrate quite clearly that many towns retained their traditional residential patterns until the second half of the century.

However, the picture changed later in the century when these towns too spawned

Plate 9.1 Short terrace of fine town houses in Victoria Place off Stow Hill, Newport, Gwent, built in the first half of the nineteenth century.

detached and semi-detached Victorian villas. The middle classes in both Cardiff and Newport were attracted to new suburbs. In Cardiff the initial outward push was as a sector to the north-east of the old town, comprising Park Place and Tredegarville; it contained impressive villas set in tree-lined avenues. In Newport there was the building of a rash of detached and semi-detached houses on the elevated sites to the north-west and east of the town. On the north-west edge they were built in an arc extending from Gold Tops (Plate 9.2) through Clytha Park to St Woollos church. Across the Usk to the east similar types of dwellings were strung out along the main roads, while the larger villas sought the sites with views in Summerhill and Maindee. This push across the river commenced in 1850 when the Maindee estate was auctioned and sold in lots, and over the next 30 years detached houses set in their own grounds were built. In 1869 the character of Maindee was described as follows:

> Some twenty years ago [Maindee] was an extensive common, but of late years has become a place of considerable importance as the site for the erection of numerous handsome villa residences, and from its contiguity to Newport, and the healthiness of the spot, has become a favourite suburban residence for the gentry and merchants of Newport. (Lewis, 1985a, 143)

Figure 2.5 in Chapter 2 presented the distribution of premises in Newport with rateable values in excess of £20 in 1880; in addition to highlighting the main commercial thoroughfares and short terraces of town houses, it shows the new peripheral suburbs in Gold Tops, Clytha Park and Maindee. So by the last quarter of the century these two ports in South Wales were showing many of the same features which had characterized the large industrial towns of England earlier in the century. Quite clearly, change occurred in different places, at different times and at different rates. As Lewis commented, 'the period and pace of residential change were not constant across the entire range of towns in the nineteenth century: while some towns were pushing out and about early in the century, others, like Cardiff, were dozing quietly' (Lewis, 1979, 150).

The beginnings of the move outwards from the commercial core was also identified by Carter and Wheatley (1982) as occurring in the second half of the century, in their study of Merthyr Tydfil between 1851 and 1871. While in 1851 the dominance of the central area in status terms was fully confirmed by the distributions of those involved in retailing, public service and the professions, of household heads in socio-economic groups 1 and 2 and of households with domestic servants, by 1871 a new middle-class residential area had emerged as a small sector to the east of the central business district in Thomastown. The retail core of the town (as measured by the distribution of household heads employed in dealing) remained, but the centre of gravity of those in public or professional occupations and of households with domestic servants had shifted towards Thomastown (see Figure 9.6). This 'out-thrusting sector of higher social status' (Carter and Wheatley, 1982, 34) was confirmed by an analysis of the addresses of those listed as 'private residents' in Wilkins's Directory of 1873: one-third were living in just two streets in this new district, Thomas Street and Courtland Terrace. Comprising not unexpectedly in a place like Merthyr relatively small town houses, this district represented the first move outwards, but it was the forerunner of what was to occur at the end of the century in the northern sector between the parklands of Cyfarthfa Castle and Penydarren House.

In addition to there being no consistency in the timing of the appearance of large scale segregated districts for the most wealthy, there are also question-marks over the social homogeneity and the timing of the development of working-class areas. Even in those towns with a very narrow industrial base and a fairly clear hierarchy of occupations, the skilled, semi-skilled and unskilled were often not

Plate 9.2 Detached villa in Gold Tops, Newport, Gwent, built in the second half of the nineteenth century.

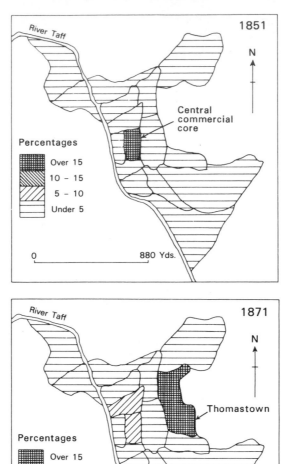

Figure 9.6 Distribution of household heads in public and professional occupations in central Merthyr Tydfil in 1851 and 1871.
(after Carter and Wheatley, 1982, Figure 20, 50)

segregated into their own discrete districts, except in the case of the very poor. The trend was towards the separation of the skilled labour aristocracy from the lower orders of workers, but as Ward (1980) and others have argued, there was considerable intermixing until relatively late in the century. Although there was a distinction between the 'rough' and 'respectable' working-class districts, the respectable areas were not the exclusive preserve of a small band of select workers. From his analysis of census data for Leeds for 1841, 1851, 1861 and 1871, Ward observed that increased segregation was characteristic of a very small proportion of the total population. He estimated that the middle class living in exclusive quarters in 1871 comprised only 2 per cent of the population, whereas the remainder of the city housed people with occupations which would be classified as middle class according to any modern taxonomy of social stratification along with the vast majority of the middling or lower middle class and the working class. The residential patterns of these three social groups and those of the skilled, semi-skilled and unskilled strata of the working class were actually less differentiated from one another in 1871 than they were in 1841 and 1851. Increased residential differentiation was restricted to the most affluent and prestigious strata of the middle class and their concern about segregation was clearly magnified by their own exceptional residential experiences. (Ward, 1980, 158–9)

He concluded that

the 'modern' residential patterns of Leeds were then created in the late decades of the nineteenth century and did not emerge from a gradual increase in the complexity of early Victorian patterns. To the degree that observers of early Victorian cities attempted to define the socio-geographic impact of industrial capitalism, they perhaps emphasized the residential segregation of the wealthy which was not new at the expense of the diminishing levels of residential differentiation amongst the less affluent. The cumulative extension of progressively higher levels of consumption to successively lower social strata which in part explains the highly differentiated housing patterns of modern cities would appear to have been of little consequence in mid-nineteenth-century Leeds. (Ward, 1980, 161–2)

Thus, while accepting the trend towards modern segregated areas, the timing of their appearance is pushed towards the end of the century.

There is general acceptance that class was one of the major factors which lay behind the eventual emergence of large-scale segregated residential areas. However, given the variety of evidence in empirical studies, it is not possible to come to any firm statement on their appearance – as Dennis wrote, 'the message of all these studies is ambiguous' (Dennis, 1984, 221). Yet, bearing in mind the fact that towns grew at different rates, at different times and in response to different stimuli, it is not to be expected that some sort of universal transformation took place in all towns at the same time in the nineteenth century. What should be stressed is that certain universal societal processes for change were active in Britain's towns and that they worked themselves out on the ground in different places, at different rates, and at different times. In other words 'Ward is probably right in questioning the view that exclusive single-class social areas had emerged with any clarity before the end of the century (the highest-status areas excepted). But it was equally apparent that the processes which were to produce such a condition were already in operation' (Carter, 1983, 199).

Ethnicity

One of the factors of differentiation which looms large in many studies of contemporary cities is ethnicity. Often this is conceived in terms of differences in race or colour between the host population and the migrant group, as in the case of West Indian and New Commonwealth immigrants in post-war Britain. However, ethnicity is much wider in scope and embraces all of those minorities who retain their cultural identities in a new environment, identities which may be expressed through language, religion or community traditions and which may or may not coincide with differences in race or colour. While colour was of no significance in British cities in the nineteenth century, ethnicity in its broadest sense was one of the factors which underpinned intra-urban spatial patterns. Yet, having made the categorical statement that it was a force leading to segregation, it is necessary to qualify it by saying that it was sometimes bound up with socio-economic status in so far as poor immigrants may have congregated together in particular districts as much through poverty as cultural cohesion.

Good illustrations of spatial concentration on grounds of religion are given in Williams's book on *The making of Manchester Jewry, 1740–1875* (1985) and in Jones's study of the Roman Catholic areas of Belfast (1960). In the Manchester example, two concentrations of Jews had emerged in the old town in the first half of the century, one at Halliwell Street, where a new synagogue had been built during 1824–25, and the other at Ainsworth's Court. Immediate access to a synagogue was, of course, a promoter of concentration. Growth in numbers led to a movement of the most affluent from these initial points out along Cheetham Hill Road and into Strangeways. At the same time as this northern axial expansion was taking place, the Jewish community was further enlarged by the arrival of poor East Europeans who settled in the Red Bank district in the 1860s and early 1870s. From 1841 to 1871 the total Jewish population increased from 625 to 3,444 and the number of families from 256 to 987. Figure 9.7 shows the distribution of Jews in 1871. Although there were still no exclusively Jewish districts in which Jews outnumbered Gentiles, and there were sharp contrasts between Red Bank and the neighbouring Anglo-Jewish extensions, the Jewish presence with its synagogues and schools gave a distinctive character to this northern sector of the city. And as Williams has pointed out, the overall effect of migrations in the nineteenth century, and particularly in the 1860s and 1870s, was to consolidate Jewish life in many other growing urban centres in the north-west, the north-east, the industrial Midlands, South Wales and the West Riding of Yorkshire. For example,

> In Leeds, where an organized congregational life had been developing very slowly since the 1820s, the migration of the 1860s took the Jewish population to over sixty families, mostly very poor, 'strict in their observance', and concentrated even then in the Leylands district. By 1866 'great numbers' of immigrants had settled in Hull. Elsewhere in the north-east, the 'back-log' effect caused by the American Civil War had given rise to thriving communities in Newcastle, West Hartlepool, North and South Shields, Sunderland and Middlesbrough. In Middlesbrough, the number of Jewish families rose from only one in 1862 to fifty (or about three hundred persons) in 1873, when the congregation built a new and larger synagogue in Bretnall Street. (Williams, 1985, 269–70)

Many of these towns have been the subjects of study in their own right, such as Krausz's examination of the history and social structure of the Leeds Jewry (Krausz, 1964). In Leeds, what Krausz describes as a 'Jewish ghetto' developed in the Leylands district, which contained 'old back-to-back houses in narrow cobbled streets and yards. The main artery was Bridge Street. In this ghetto lived the bulk of the Jews, but just outside it to the west was Belgrave Street and its environs where the main synagogue was situated and where some of the elite resided' (Krausz, 1964, 21). Most of the immigrants were Orthodox Jews and their community was sustained by the synagogue, the schoolroom, the ritual bath and the kosher butcher shops. In addition, their poor command of English and some open hostility by the host population further consolidated their community spirit and separateness. In due course, as the Jews became assimilated through more frequent contact with the non-Jewish population and as their wealth increased, they moved out to the northern districts of Chapeltown Road, Moortown and Alwoodley, but this did not get underway until the early decades of the twentieth century. This process of suburbanization of Jews and their tendency to move outwards in a straight line or a series of straight lines has been discussed by Lipman (1968), who has also provided

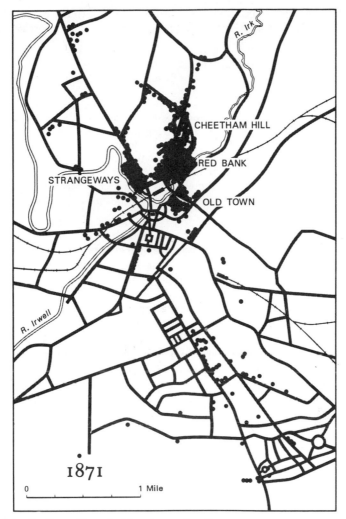

Figure 9.7 Distribution of Jews in Manchester in 1871.
(after Williams, 1985, map 8, 369)

a detailed description of the Jewish experience in London. What all of these studies show is that in the initial colonization and subsequent dispersion of Jews, religion played a central role in sustaining distinctive communities in many British cities.

The second and particularly striking example of spatial segregation on grounds of religion is provided by Catholicism in Belfast (Jones, 1960). Before the nineteenth century there is no evidence of conscious segregation between Catholics and Protestants, but consequent upon the large influx of Catholics in the first half of the century to the job opportunities in the growing industrial town, a clear separation between the two groups emerged. Whereas in the mid eighteenth century the proportion of Catholics was only some 6.5 per cent, by 1861 it had increased to 34.1 per cent. The result was that tension between the two communities rose and rioting followed, which further heightened spatial segregation. Jones notes that this segregation became more and more rigid, to the extent that distinct districts could be demarcated on the ground, as shown in a report on the riots of 1886:

> The extremity to which party and religious feeling has grown in Belfast is shown strikingly by the fact that the people of the artisan and labouring classes . . . dwell to a large extent in separate quarters, each of which is given up almost entirely to persons of one particular faith, and the boundaries of which are sharply defined. In the district of West Belfast, the great thoroughfare of Shankill Road, with the network of streets running into it, and the side streets connecting those lateral branches is an almost entirely protestant district . . . the great catholic quarter is due south of the Shankill district, and consists of the thoroughfare known as the Falls Road, and the streets running south of it. Due south of the Falls district is Grosvenor Street, almost entirely inhabited by protestants, so that the catholic quarter lies between two protestant districts. (Jones, 1960, 191)

Religion was also of importance in binding together Welsh migrants in English cities, but there was for many the additional cohesion afforded by the Welsh language. In the middle of the century both Liverpool and London had Welsh communities of more than 10,000 persons born in Wales, and Shrewsbury, Chester, Manchester, Bristol and Birmingham had more than 1,000 Welsh-born. From evidence for Liverpool in particular, Pooley has shown that the Welsh sought to preserve elements of their culture in the larger cities. However, in the case of Liverpool, the Welsh were not tightly clustered but could be found in most areas (Figure 9.8). While Everton could be described as the main Welsh district, there were many small foci across the city. In terms of religion, it was the nonconformist chapel which acted as the centre for communal activities.

> By 1871 in Liverpool 23 Welsh chapels had been established, located mainly in those areas containing the largest concentrations of Welsh population, and drawing their exclusively Welsh congregations from almost all parts of the Welsh community. Evidence suggests that even those Welsh migrants who lived some distance from the main Welsh residential areas were prepared to travel to chapel, and a network of chapels, Sunday schools, and other educational and social organizations, all mainly conducted in the Welsh tongue, served to provide both a focus for the Welsh community and also a medium through which traditional Welsh values and the social hierarchy of rural Wales could be maintained in an urban environment. (Pooley, 1983, 299)

What is apparent is that attendance at the Welsh chapel meant more than religious worship: it was a meeting place where the Welsh language could be used and where news from home could be spread and discussed, so religion and language interacted to maintain cultural cohesion. On top of this there were other activities which helped to sustain their identity.

> Welsh affairs were also of sufficient importance to merit the circulation of at least two Welsh-language newspapers in Liverpool in the nineteenth century, whilst the English-language *Liverpool Mercury* also gave extensive coverage to Welsh news, thus helping to bind the community together. In addition there was a network of local clubs and *eisteddfodau* which served the working-class Welsh community, together with the more formal Welsh National Eisteddfod held in Liverpool in 1840, 1884, 1900 and 1929, and the Welsh National Society, formed in Liverpool in 1883, to serve the increasing number of middle- and upper-class Welsh businessmen and professionals in the city. (Pooley, 1983, 299)

Thus, in the case of the Welsh, although there was no tight spatial clustering, there were a number of living communities with a very distinctive cultural character. This

Figure 9.8 Residential distribution of Welsh-born household heads in Liverpool in 1871. (after Pooley, 1983, Figure 7, 300)

example also highlights the point that religion, language and community activities often interacted together to determine the distinctive personalities of ethnic communities.

In the cases examined above there is evidence that religion and language did foster cohesion among certain immigrant groups. However, in other cases it is less easy to determine whether it was ethnicity which drew people together or whether it was economic necessity which forced them into certain residential districts. One group to which this applies is the Irish who settled in large numbers in many industrial and port towns. Their spatial concentration in the middle decades of the century has been well documented in numerous empirical studies, from which the conclusion can be drawn that they occupied the poorest residential quarters, at least until they found their feet in the new urban environment. With few skills and possessions on arrival it is not surprising that they were relegated to the bottom of the housing market and that they took on the most menial of labouring jobs or the pettiest of trades. Typical examples of their spatial clustering are given in studies of Liverpool (Pooley, 1977) and Cardiff (Lewis, 1980). In the former, 'the Irish core area was clearly in north-central Liverpool, an area of mainly high-density, sub-standard housing, which formed a low-status zone' (Pooley, 1977, 370). In Pooley's correlation analysis of housing variables, the Irish in 1871 were strongly associated with multiple-occupancy, courts, and high-density sub-standard property. Cardiff also had a sizeable Irish population by mid-century, a community which comprised about 12 to 15 per cent of the total population of the town in 1851. Lewis has shown that they were concentrated in the poorest housing, that is, in the congested courts behind the main streets in the medieval nucleus, such as Landore Court and Kenton's Court, and in the newly constructed terraces of small houses to the south-east of the old town, particularly in Stanley Street, David Street, Mary Ann Street, Love Lane, Little Frederick Street and Newtown (Figure 9.9). This pattern was confirmed in Paine's Public Health Report for 1855: the West 'has many courts in a bad sanitary condition, as Landore Court, Mill Lane Court, Evans's Court and others. They are occupied by labouring and indigent Irish, who live in common lodging-houses'. In the east 'some of the streets – as Stanley Street, Love Lane, and Mary Ann Street – contain Irish lodgings-houses'. He observed that in Newtown 'the houses for the most part are occupied as Irish lodging-houses, and are seriously overcrowded' (Lewis, 1980, 21). Lewis has shown that these districts were characterized by overcrowding and poor sanitation, and that mortality and

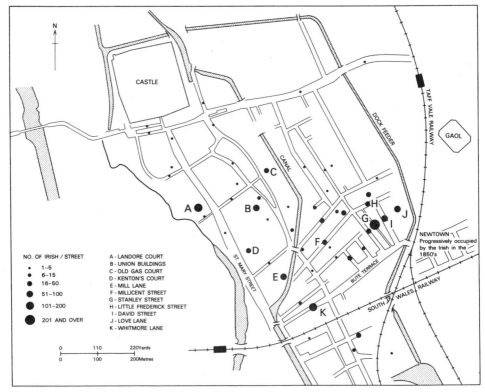

Figure 9.9 Distribution of Irish in Cardiff in 1851.
(after Lewis, 1980, Figure 3, 22)

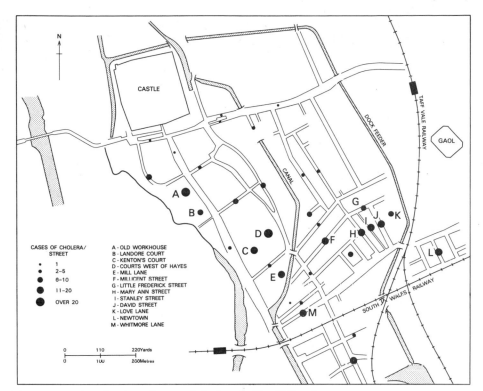

Figure 9.10 Distribution of cases of cholera in Cardiff in 1849.
(after Lewis, 1980, Figure 8, 37)

disease rates were very high. Figure 9.10, which maps the distribution of cases of cholera in 1849, matches closely the distribution of Irish in Figure 9.9. This concentration into areas of poor housing has been corroborated in studies of many other towns, such as Huddersfield (Dennis, 1984), Leeds (Dillion, 1974–79), Bradford (Richardson, 1971) and Merthyr Tydfil (Wheatley, 1983).

At the cultural level, it can be argued that a sense of community may have been engendered through the Catholic church (or for that matter through Protestant worship since not all of the Irish were Catholics). In the case of Cardiff, a Catholic church and Catholic school were built between Stanley Street and David Street right in the heart of the main Irish quarter and served as focal points. Yet it is difficult to be precise about the strength of religion in welding the Irish together as a community. For the very large Irish population in Liverpool Pooley found little evidence to suggest that exclusively Catholic-Irish societies or institutions were of great significance (in marked contrast to the situation in many cities in North America). He went on to speculate that perhaps Irish cultural cohesion did not manifest itself in formal structures, rather 'in the working-class Irish areas such cultural coherence was centred around informal associations – perhaps based on the corner pub – which leave no record, but which may be no less important in their effect on the community' (Pooley, 1977, 378).

If the role of religion as a powerful cohesive force amongst the Irish is open to question, there is little dispute over their poverty, and this has been widely used to explain their gravitation to the poorest housing. In Liverpool, the Irish were strongly associated with low social status (social classes 4 and 5), and in Cardiff some 16.86 per cent were in class 4 and 66.15 per cent in class 5, the latter reflecting the large numbers described as general labourers, charwomen and hawkers in the census schedules. Their generally low status and their concentration in specific districts was also confirmed by Wheatley (1983) in Merthyr Tydfil. In her analysis of data abstracted from the 1851 census, various measures of low status were strongly associated with those of Irish birth – labourers, social classes 4 and 5, and multiple occupancy – and the majority were living in an area known locally as 'China' which 'could almost be called an Irish "ghetto"' (Wheatley, 1983, 331). Contemporary descriptions show quite clearly that they were at the bottom of the economic hierarchy in this iron-making town. The *Morning Chronicle* in 1850, for example, noted that

> It is a remarkable fact that the Irish, no matter what length of time they remained at the works, never rise to the dignity of skilled workmen. Such a wonder as an Irish puddler was never heard of. They have the same opportunity as others; and, even supposing the jealousy and antipathy of the Welsh in some degree to prejudice them, by patience and good conduct they would surmount those obstacles, and take their place in the front ranks of the workmen; yet, whatever be the cause, they never do so, but remain contentedly in the heavy drudgery to which they are at first admitted. (Wheatley, 1983, 133)

Later in the decade it was recorded that 'the Irish . . . seldom rise to the rank of skilled or highly paid labourers' (Wheatley, 1983, 133). These studies suggest that the main factor which accounted for their concentration was their low socio-economic status. Pooley's conclusion on the Irish in Liverpool was unequivocal: 'socio-economic status was undoubtedly the most important non-cultural characteristic which affected residential location, and the uniformly low social status of many Irish migrants went a considerable way towards explaining the Irish concentrations' (Pooley, 1977, 378).

In addition to the cultural and economic reasons for segregation it can also be argued that hostility towards the Irish on the part of the host population may have encouraged a degree of spatial clustering. There is a lot of evidence that the Irish were seen as undesirable newcomers for a variety of reasons, including cheap labour, disease, drunkenness, petty crime and dirty habits. The words of Engels, who drew on the observations of Thomas Carlyle and Dr Kay, illustrate the extreme reaction which came from some quarters:

> . . . the Irish have, as Dr Kay says, discovered the minimum of the necessities of life, and are now making the English workers acquainted with it. Filth and drunkenness, too, they have brought with them. The lack of cleanliness, which is not so injurious in the country, where population is scattered, and which is the Irishman's second nature, becomes terrifying and gravely dangerous through its concentration here in the great cities. The Milesian deposits all garbage and filth before his house door here, as he was accustomed to do at home, and so accumulates the pools and dirt-heaps which disfigure the working-people's quarters and poison the air. He builds a pig-sty against the house wall as he did at home, and if he is prevented from doing this, he lets the pig sleep in the room within himself . . . Drink is the only thing which makes the Irishman's life worth having, drink and his cheery care-free temperament; so he revels in drink to the point of the most bestial drunkenness. The southern facile character of the Irishman, his crudity, which places him but little above the savage, his contempt for all humane enjoyments, in which his very crudeness makes him incapable of sharing, his filth and poverty all favour drunkenness. (Engels, 1969, 124–5)

Contemporary descriptions and health reports are full of cases where the Irish were blamed for almost all the ills of society. In Cardiff, for example, they were castigated for being the harbingers of disease – 'these straggling accessions' to the town 'generally consist of the most wretched members of the society from which they have, as it were, been cast forth – generally in a starving condition, often already afflicted with disease, or carrying the seeds of it about them. Of necessity many of these immigrants will die before they have dwelt so long in the new community upon which they have thrown themselves . . . Their afflictions, moreover, will extend to their neighbours; and contagious diseases imported by them will materially modify the sanitary condition of the place' (Rammell, 1850, 14). Then again, as Richardson observed in Bradford, 'their readiness to fight, and drunkenness, may go some way towards explaining why the Irish were not acceptable as neighbours and the consequent tendency for the non-Irish to move out of the Irish streets, their place being taken by Irish' (Richardson, 1971, 312). In passing, it should be noted that hostility was not confined to the Irish, but sometimes surfaced with other immigrant groups as well and may have contributed to their segregation. In Krausz's study of the Leeds Jewry the point was made that 'the Jews of the Leylands were even exposed to violence. The district was a slum area, the haunt of the underworld and a hotbed of drunkenness and immorality, which bred bands of hooligans who spread terror amongst all peacefully minded citizens, although it was the Jews who suffered most at their hands' (Krausz, 1964, 21–2). What is clear from the discussion of the Irish is that segregation cannot be simply ascribed to ethnicity. While cultural cohesion may have played a part in their clustering, it can be shown that their lowly socio-economic status was of greater significance. Further, there was some hostility by the hosts which may have encouraged spatial concentration. To an extent, this complex of cultural, ecomomic and social causes may have applied to other groups as well.

In conclusion to this consideration of ethnicity, attention can be drawn to Pooley's useful attempt, based on his studies in Liverpool, to summarize in a diagram the interaction which occurred between economic, socio-demographic and cultural factors (Pooley, 1977, 378–80). In his formulation (Figure 9.11) he suggested four extreme theoretical situations in which migrants could find themselves in the city: first, that of total assimilation and residential integration, where the migrant had the same economic, socio-demographic and cultural profile as the host population (such as the majority of English migrants); second, that of distinctiveness by virtue of pronounced demographic or economic characteristics, but showing no strong cultural traits – he provided no hard-and-fast example of this but observed 'micro-level variations in the location of migrants from different parts of England probably fall into this category, and the available evidence suggests that socio-economic factors may have been more important than cultural factors in causing the segregation of Irish migrants on first arrival in the city' (Pooley, 1977, 378); third, that of distinctiveness by virtue of strong cultural cohesion, but showing no pronounced demographic or economic features (it was the Welsh who fell into this category); and, finally, that of distinctiveness on both socio-economic and cultural counts (and it was the Irish who occupied this extreme 'ghetto' situation with low socio-economic status, general lack of urban experience, cultural insularity and a degree of discrimination). What is particularly useful about this diagram is that it reminds us that while ethnicity was often closely bound up with economic considerations, particular groups did stand out to a greater or lesser extent on the strength of their cultural cohesion, at least in the early stages of their settlement in towns.

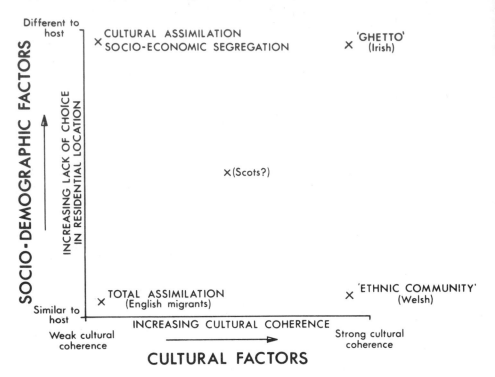

Figure 9.11 Diagram showing the interaction of cultural and socio-demographic factors affecting the residential segregation of migrant communities. (after Pooley, 1977, Figure 8, 379)

Mobility

Finally in this consideration of the key factors which lay behind residential segregation, attention must be given to mobility. Two aspects in particular need brief examination, the improvements in transport technology which in part followed and in part encouraged the outward spread of cities, and the intra-urban patterns of mobility by which people of like class or ethnic background came together into their own distinctive residential districts.

In terms of improvements in technology, the century witnessed a transformation from 'the walking city' to the city served by the electric tramcar. Over the course of the century new forms of transport facilitated cheaper and longer-distance commuting, and the greater separation of residence and workplace for some sectors of the population. There were those, though – particularly the less skilled of the working class who were in casual employment – who continued to demand residences near to their places of work. A useful summary of the main changes which occurred has been provided by Dennis (1984) and this will be used here. The wealthy who first moved to the suburbs early in the century usually had their own private carriages; they could live in their preferred locations and could reach the city centre or their business premises in their own conveyances. From the 1830s onwards horse omnibuses were introduced by enterprising operators to these growing suburbs, but as Dennis has commented, 'nowhere was the omnibus either essential or pioneering' (Dennis, 1984, 117). As the less wealthy middle classes turned to the suburbs further routes were opened up; they were generally restricted in their clientele for their high fares and timing meant that they were little used by the working-classes. In the 1870s the horse omnibus was supplanted or supplemented by horse-drawn trams, but they also appeared initially on established middle-class omnibus routes. In Leeds, for example, the first five routes authorized under the Tramway Order of 1871 served the same destinations as the first five bus routes some thirty years or so before, and in Huddersfield, too, the first lines replicated bus routes. However, the last decade of the century saw the use of electric traction on the tramways and it was this which had the effect of reducing fares, and for the first time attracting a wider working-class custom. Thus for much of the century road-passenger transport largely catered for the middle classes, but with electrification 'at last services were both cheap enough and conveniently timed to attract working-class patronage, for commuting and recreation. Contemporaries clearly believed that the electric tramcar was the "gondola of the people" and the key to suburban growth' (Dennis, 1984, 123–4). Although electric trams may not have been of great benefit for the poorest workers living in inner districts, they certainly helped the better-off working classes to espouse suburban living, and benefited the middle classes who had moved to the outer suburbs. While it was road transport which had most influence in individual towns across the country, in specific cases, particularly the large cities, the railway had some impact on suburban spread. In general, though, the railway companies were wary of what were often loss-making suburban routes which suffered intense competition from tramways.

At the same time as improvements in transport technology enabled workers to overcome the friction of distance for commuting, city dwellers themselves sought to find a residential area which best matched their perceived (and financial) status in society, an area which at least suggested the qualities of the higher classes and which was clearly separated from the lower classes. In effect, there was a sifting and sorting process taking place as residents moved around the city in search of their spatial niches. While keeping one eye on their place of work and access to it, for many the other eye was on finding a suitable location where they could display their social standing. The outcome of this internal mobility was the emergence of the districts segregated by class discussed earlier in this chapter.

Pooley (1979) has shown that it is not a straightforward task to reconstruct intra-urban patterns of mobility because all of the potential data sources – the census, directories, rate books, electoral rolls, personal papers – present problems in use. However, by painstaking analysis of this material, a number of writers have derived data on both mobility and persistence rates. Of these studies, that by Pooley on Liverpool is one of the more robust, from which he has drawn some useful generalizations which have been incorporated in a model of the residential mobility process (Figure 9.12). Most of Pooley's data were derived from the linking of census and directory sources through two decades, 1851–61 and 1871–81. For the different social areas identified in the Liverpool of 1871, a sample of 2,446 households was drawn for detailed study in each decade. Making due allowance for mortality, and the inevitable impossibility of tracing some of the sample over time, he was able to examine persistence and mobility, and to relate the patterns to the social, economic and demographic characteristics of households. Pooley's work is also interesting for its use of personal records, in particular the diary of David Brindley, to add some detail and colour to the bare data. This diary illustrates both the strengths and weaknesses of such personal papers: while providing vital information on individual moves, they are relatively rare in record offices, and may not shed light on the precise reasons for moves. Brindley's diary, which covered the period 1856–91, but is best for the 1880s, allowed the reconstruction of the many short moves of this 'respectable' working-class family man in the Everton-Kirkdale area of the city (see Table 9.3). While revealing the spatial patterns, the motives for moving both to Liverpool in the first place and within the city were not really spelled out. For example, in a separate paper on this source by Lawton and Pooley, the reasons for residing in Everton-Kirkdale had to be inferred: 'the need for close contact with relatives and friends must have been a major constraint on movement, as must have been distance from work' (Lawton and Pooley, 1974, 154). Nonetheless, this material does add further substance to Pooley's observations on the frequency, distance and direction of moves.

In Pooley's assessment, the residential mobility process incorporates three major elements: the personal circumstances and housing conditions which affect movers, the distance, direction and frequency of moves, and the consequent creation of different types of residential areas within the city. Evidence on the relationships between these three elements in Liverpool enabled him to suggest the tentative model of the process and its effect on urban structure shown in Figure 9.12. It can be seen that the spatial outcome was a variety of residential areas differentiated by status, and showing varying degrees of stability, ranging from community formation at the one extreme to disintegration at the other. While improvements in

Table 9.3 Residential moves of David Brindley, 1882–1890 (after Lawton and Pooley, 1974, Table 1, 154)

Move No.	Address	Distance moved (metres)	Duration of residence (days)
1	Milkhouse 1 Brookhill Road, Bootle		
2	Milkhouse 103–105 Woodville Terrace, Everton	4,200	206
3*	22 Hamilton Road, Everton	800	36
4*	42 Langham Street, Everton	1,650	Approx. 3.25 years
5	68 Everton Brow, Everton	2,300	Approx. 163 days
6	56 Landseer Road, Everton	600	79
7	13 Wentworth Street, Everton	140	99
8	14 Mere Lane, Everton	500	199
9	64 Rumney Road, Kirkdale	1,500	14
10	29 Norgate Street, Everton	960	93
11	22 Norgate Street, Everton	25	198
12	8 Downing Street, Everton	950	798

Note: * Information for moves 3 and 4 is inconclusive.

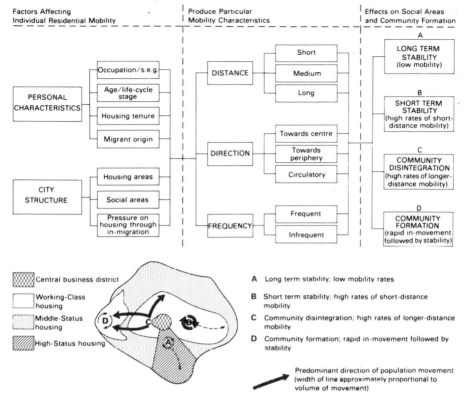

Figure 9.12 The residential mobility process and its effect on nineteenth-century urban structure.
(after Pooley, 1979, Figure 6, 274)

transport technology progressively reduced the friction of distance and facilitated centrifugal spread, differential levels of mobility underpinned the formation of segregated districts within the growing city. In Pooley's own words

> Whatever the precise form of mobility patterns, it is clear that the Victorian city formed a dynamic system, where individual residential moves were frequent, but which collectively provided a sorting mechanism, reinforcing some communities and destroying others, and allowing a transition to take place from the essentially pre-industrial town of the early nineteenth century to the well-structured industrial city of the mid- and late-Victorian period. (Pooley, 1979, 275)

As noted earlier, perhaps not all cities showed the well-structured modern form by the middle of the century, but at least the dominant trend was in that direction.

Conclusion

The conclusion to this chapter has been anticipated in the spatial models presented in Figures 9.1B, 9.2 and 9.12: the demand for urban living space was met through the emergence of a variety of segregated residential districts which became the 'hallmarks' of the modern industrial city. However, segregation itself was not new – this had been a feature of all towns before the nineteenth century. What changed was the scale of segregation. In the early 1800s, towns still contained elements of their pre-industrial segregated structures, with pronounced front street/back street contrasts, and perhaps contrasts between floors under the same roof as well. In brief, segregation was there, but it was small-scale. This pattern remained in the central parts of many towns until well into the century. It has been shown in this chapter that the evolving class structure, ethnicity and new levels of mobility were powerful forces which bore down on the internal composition of cities and which, in combination, led to the appearance of large-scale segregated districts. First the wealthy were able to detach themselves from the urban core and move to suburban enclaves. This set the trend for the middle classes in general. And in due course, members of the labour aristocracy began to separate themselves out from the mass of the working class and move into their own 'elite' districts. Superimposed on top of this was the distinctive character given to certain areas by ethnicity, as expressed through language, religion and common background.

Taking into account the conclusions drawn in Chapter 6 and 7, that the city also acquired specialized commercial and industrial districts, it is possible to offer a simple model of the internal structure of the evolving modern city which shows how the competing demands for urban space were met (Figure 9.13). The main elements in this are:

1 an emergent central business district. This became dominated by commercial activities as middle-class shopkeepers, businessmen and professionals moved to suburban residences.
2 an inner mixed zone, comprising slum property, often housing ethnic minorities, and miscellaneous commercial activities.

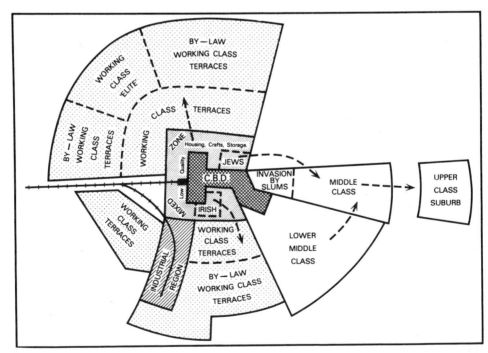

Figure 9.13 Diagrammatic model of the Victorian city.
(after Carter and Lewis, 1983, Figure 22, 90)

3   sectors of middle-class housing pushing out to suburbia.
4   extensive tracts of working-class terraces, sometimes associated with an indust-
    rial region; the improved terraces of the late nineteenth century were subject to
    building regulations, as will be discussed in Chapter 12.

What must be stressed about this model is that it 'is not a model for *the*
nineteenth-century city, for there was no such standard entity – different cities
responded to their own particular varieties of industrialization at different rates
and at different times' (Carter and Lewis, 1983, 90). However, what can be
suggested is that during the century towns began to display these modern segregated residential and other districts in response to the processes which have been
discussed in this part of the book.

10

The supply of urban space

Having looked at the composition of those who generated the demand for urban accommodation, this chapter and the next will examine the response of the property market to meet that demand. In effect, this means looking at the activities of all those involved in urban development, from the initial release of land through to the completion of new houses and other premises. Many writers who have examined the structure of the development process have adopted a decision-making approach to try to identify the contributions of various individuals to the eventual built form. For example, this was the approach set out by Powell in the opening sentences of his recent book: 'Every building ever built arose from a carefully premeditated decision which was usually of the utmost importance to the person who made it. Accordingly, we begin at the beginning by looking at the people who decided to build, who they were, why they built, and how they set about the task' (Powell, 1982, 1). Similarly, in her study of land development in Huddersfield in the nineteenth century, Springett favoured a decision-making framework. She worked from the premise that 'the mechanism of land development is a decision-making process that involves a number of decisions taken by one or more key decision-makers, which governs the evolution of land from one state to another' (Springett, 1979, 24). It is this approach which will be followed here.

A useful starting point is Springett's attempt to summarize the decision-making process in a flow diagram, showing the links between a variety of primary and secondary decision makers (Figure 10.1). The central figures were the landowners and builders, although they may have been linked by an intermediary, a developer, who provided some basic facilities on site, such as roads, pavements and drains. In turn, they were serviced by secondary professionals – solicitors, land agents, estate agents, land surveyors, bankers – who dealt with legal, technical and financial matters, and by a veritable army of small capitalists who put up the mortgages for building operations. This diagram has the obvious advantage of clarity, but it should be noted that it does over-simplify the roles of the various figures by placing them in separate boxes. It was not uncommon for an individual to play more than one role, such as the landowner who engaged contractors to lay out roads on his land and then went on to provide the capital for builders to complete their schemes, or the shopkeeper who purchased land and then acted as a builder by employing craftsmen to build houses on his site. As Powell observed, 'landowner, sponsor, developer and builder may have been different or one and the same and, if different, their respective interests may have coincided or conflicted' (Powell, 1980, 1). Perhaps, also, the diagram implies that the working of the building industry is well

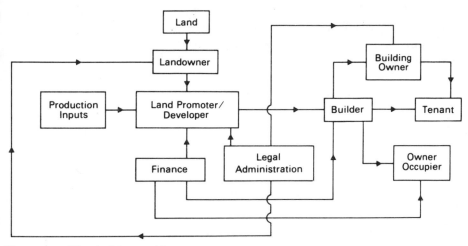

Figure 10.1 The decision-making process.
(after Springett, 1979, Figure 1.1)

known in all its detail when, in reality, there are gaps in our knowledge, particularly about those very small-scale speculators, developers and builders 'whose identity and motives often remain shadowy' (Powell, 1982, 1). Nonetheless, bearing these limitations in mind, this decision-making framework will be employed to examine the supply of urban land and property. What is of central concern is the influence or imprint of those involved on the allocation of land to the varied urban uses and on the eventual built form. The discussion will be arranged around the primary figures of landowners in this chapter and builders in Chapter 11. While particular attention will be given to their activities in the residential sector, it should be remembered that their influence extended to the other urban uses discussed in Chapters 6, 7 and 8.

Landownership

The word landowner encompasses a tremendous variety of individuals and groups who held parcels of land which were ripe for urban development. At the one extreme were the titled gentry who owned extensive estates in and around Britain's expanding towns. At the other extreme, land was fragmented into very small building plots which were in the hands of a vast number of small owners. While most land was in the private ownership of individual families, some companies and corporate bodies, such as charities, hospitals, schools and universities, were significant landowners in particular localities. Thus, in all but the most exceptional of cases, the conversion of land to urban use could involve the decisions of numerous proprietors or their representatives.

There are two particular aspects of landownership which warrant attention, first, the influence of the owners themselves on urban form, and, second, the influence of the spatial configuration of the underlying land holdings on subsequent layout. The landowner could exercise a degree of control over the direction and pace of urban expansion by his willingness or unwillingness to release land for development, and he could partly determine what was built on his property by his own participation in its development or by the inclusion of restrictive clauses in the leases or other terms granted to builders. Beyond that, the shapes and sizes of the land holdings which were released could dictate the layout of streets and houses. In so many cases, individual building schemes had to be tailored to fit into the awkward shapes of pre-existing plots. These two aspects of landownership will be discussed in turn.

Attitudes to development

The reactions of owners to demands for building land were varied, ranging from outright refusal to positive encouragement with financial backing. On the negative side, there were those who were unwilling to turn over their fields to builders. For the majority it was a short-lived refusal; some, though, were prepared to hang on to their 'rural' enclaves, even when encircled by bricks and mortar, not necessarily for their agricultural worth but perhaps to maintain personal privacy or in anticipation of mineral wealth. Whatever the reason, the negative decision of a landowner could act as a brake on the extent and direction of urban spread. At the same time, the resistance of unco-operative owners could have the effect of putting even greater pressure on building ground in a town's nucleus, and on those plots which did come onto the market.

One of the classic, well-documented examples of opposition to suburban expansion was Nottingham, where for a long while the landowners on its fringe successfully resisted the demand for new building ground and thereby intensified the pressure on its core. At the beginning of the eighteenth century, Nottingham's 10,000 inhabitants lived in fairly spacious surroundings, largely within the circuit of the medieval town. However, in the second half of the century, with the industrialization of its knitwear and lace production, its population increased enormously and was housed at very high densities within the 800 acres of the medieval core. By 1831 its population had reached over 50,000, and the old town was heavily congested: '. . . narrow streets and alleys lined with miserable hovels were made between the existing thoroughfares and every interior space was filled with dwellings and workshops. Squalor was rampant' (Edwards, 1966, 370). The primary reason for this state of virtual imprisonment was the persistent opposition of the burgesses and freemen to the enclosure of the common lands which lay around the town. These lands comprised Sand Field and Clay Field to the north, and East Croft, West Croft and the Meadows on the Trent flood-plain to the south (Figure 10.2). Altogether about 1,400 acres of common land lay within the municipal boundary but outside the built-up area. For the first four decades of the nineteenth century, those opposed to enclosure were able to hold out against the demands for building land through their control of the Charter Corporation, but with the election of a new broader-based Borough Council following the Municipal Reform Act of 1832, the way was opened up to end 'the restrictive feudal tenure of the circumjacent commonable lands' (Gray, 1953, 66). In the first instance, by Enclosure Acts of 1839, some 52 acres in West Croft, Burton Leys and Lammas fields were allocated to claimants. Shortly afterwards, by the Nottingham Enclosure Act of 1845, this process was extended to the rest of the commons, and over the next 20 years the enclosure commissioners '. . . set out allotments according to the claims of the various owners, of whom there were about 400. They had to arrange the allotments according to the road plan which seems generally to have followed the pre-enclosure field paths; and these go back to the Middle Ages and beyond . . . Within the pattern thus laid down by history or by nature the allotments of all shapes and sizes were set out, each in the possession of a separate owner. This system was followed over the entire area of the Meadows and the Sand Field and Clay Field' (Chambers, 1952, 11–13). Thus in the case of Nottingham the

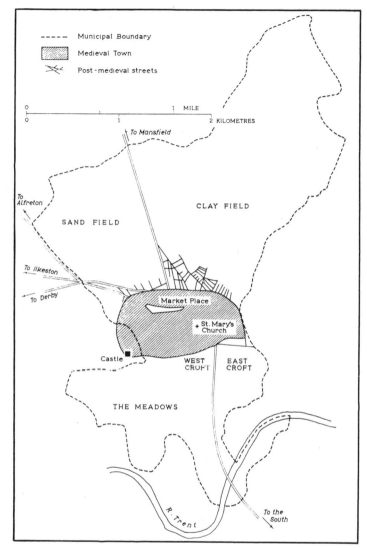

Figure 10.2 The common lands of Nottingham in the fifteenth century.
(after Edwards, 1966, Figure 63, 369)

resistance to building was eventually broken, but not before it had had a major impact on the internal configuration of the city.

Nottingham was a rather extreme case, but it illustrates well the impact of recalcitrant owners on the shape and spread of towns. Elsewhere the opposition may have been less striking, but nonetheless a factor to be taken into account. In Cardiff, for example, the Bute family was generally sympathetic to requests for building land, but it preserved a large park adjacent to its residence, the castle, near the core of the medieval town. To this day the castle grounds, Sophia Gardens and Cathays Park stand out as a very distinctive green swath right at the heart of the modern city. Slater has shown that it was quite common for the landscape parks of the gentry to act as buffers to development on the edges of towns for part, if not all, of the nineteenth century (Slater, 1977). Indeed, it was not beyond the power of

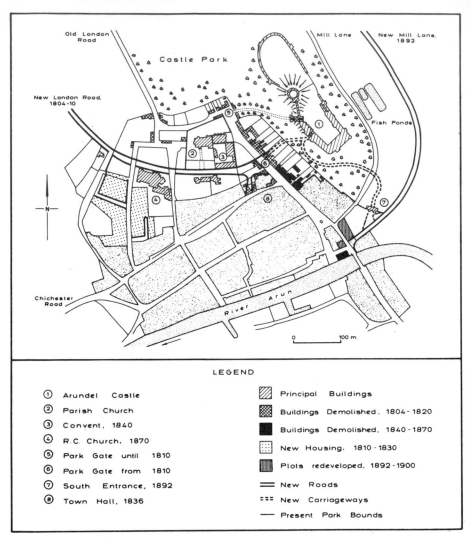

Figure 10.3 Arundel: nineteenth-century change in town plan.
(after Slater, 1977, Figure 2, 322)

landowners to extend their parks into towns by demolishing cottages and shops and rearranging the road layout. One of Slater's examples, Arundel, shows the way in which a landowner, in this case the Duke of Norfolk, protected the privacy of his residence. In the late eighteenth century the land around the castle was laid out as a park, but the town itself was uncomfortably close to the castle along the Old London Road leading into High Street. As can be seen from Figure 10.3, early in the nineteenth century the Duke demolished property and altered the road layout: 'the new London road entered the town through gardens and orchards south of the parish church. This necessitated the demolition of houses only where it broke through into High Street. However, the eleventh Duke of Norfolk took the opportunity this scheme presented to close all the roads north of the new turnpike, to demolish the cottages lining them, and to enclose the whole area, except the church, into the park. The displaced cottagers were rehoused by the Duke, on a newly constructed estate of well-built houses in the western part of the town' (Slater, 1977, 321). Subsequently, between about 1840 and 1870, further properties which overlooked the castle on the east side of High Street were demolished, and 'in the 1890s, after protracted arguments with the town Corporation, the fifteenth Duke of Norfolk was allowed to realign the road to South Stoke which ran below the eastern ramparts of the castle' (Slater, 1977, 322). Thus successive Dukes of Norfolk were able to steer urban development away from their castle residence. This sort of action, of course, was simply a continuation of what had occurred widely in the eighteenth century as landowners sought to improve the vistas from, and the privacy of their homes. Some owners went as far as razing entire settlements

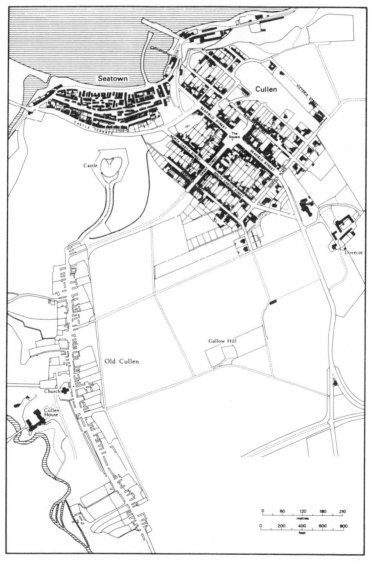

Figure 10.4 The lost burgh of old Cullen and its replacement, the new town of Cullen, commenced in 1822.
(after Adams, 1978, Figure 3.13, 70)

and replacing them by new villages or towns away from the immediate environs of their mansions, for example, the Earl of Dorchester at Milton Abbas in Dorset and the Earl of Seafield at Cullen (Figure 10.4) and the Duke of Gordon at Fochabers in Scotland (Adams, 1978, 68).

However, on the positive side, there were those who were ready for the builders to move in. In fact, some who protected the parkland around their own homes were quite willing to entertain development on estates elsewhere, as for example, the involvement of the Dukes of Norfolk in the development of Littlehampton just a short distance from their Arundel residence. For many the conversion of agricultural land to urban use was a major source of revenue. In London, where according to a mid-century commentator, land was 'far more valuable than that round any town in the world', a long list of large owners benefited from its development – the Dukes of Bedford, Portland and Westminster, the Lords Berkeley, Portman and Southampton, the Cadogan (Chelsea), de Crespigny (Camberwell and Peckham), Evelyn (Deptford), Ladbrooke (Kensington) and Fox (Kensington) families, and so on (Spring, 1971, 39–40). On the Bedford estate, for example, from the 1830s onwards the Dukes received an impressive rental in excess of £70,000 per annum, partly from ground rent and partly from rack rent. As explained by an agent, the estate could count on a steady and permanent revenue:

> The old parts of the Estate [he wrote] sink into ground rents, thro' the houses upon them requiring to be rebuilt, the new parts will rise by the falling in of the building leases, from ground-rents to house-rents, and this process will always be going on as the new parts of the Estate became old and the old parts new: and thus kept up, the Rental . . . may be said to be as permanent as if it were the Rental of a Landed Estate (Spring, 1971, 41).

Outside London, the same pattern was repeated in and around Britain's towns as owners or their agents agreed terms with builders. In some instances the owner abandoned his family seat and moved elsewhere, before realizing the building potential of his vacated estate, such as occurred in Edgbaston, Birmingham, when Lord Calthorpe moved from Edgbaston Hall to Ampton in Suffolk, and in West Bromwich when Lord Dartmouth moved from Sandwell Hall to the Patshull estate. A brief indication of activity by landed families in provincial England is given in the following extract taken from one of Spring's studies of English landowners:

> In turning to provincial England, it soon becomes apparent that many sizeable towns had one or more landed families receiving income from ground and houses. In the South there was the Duke of Devonshire at Eastbourne and Lord Radnor at Folkestone; in the South-West, the Marquess of Ailesbury at Marlborough, the Duke of Cleveland at Bath, Lord St Levans at Devonport, the Palks at Torquay and Lord Clinton at Redruth. In the Midlands and the North, there were Lords Calthorpe, Dartmouth and Hertford in Birmingham, the Duke of Cleveland in Wolverhampton, the Marquess of Westminster in Chester, the Earl of Wilton in Manchester, and in Liverpool Lord Stanley of Alderley, the Marquess of Salisbury and the Earls of Sefton and Derby; the last, it may be noted, also held property in Macclesfield, Bury and Preston. In the North-East there were the Duke of Northumberland on Tyneside, the Earl of Durham at Sunderland, the Marquess of Londonderry at Seaham, and the Ridleys at Blyth. In Yorkshire there were the Ramsdens at Huddersfield, Lord Howard of Effingham at Rotherham, Lord Rosse in Bradford, and Earl Fitzwilliam and the Duke of Norfolk in Sheffield. Finally, in Lincolnshire and its neighbourhood, there were the Earl of Yarborough and the Heneage family at Grimsby, Lord Burleigh at Stamford, Earl Fitzwilliam at Peterborough, and the Duke of Newcastle in Nottingham. (Spring, 1971, 42–3).

Of course, to this list of notables can be added the very large number of small owners who turned over their parcels of land to bricks and mortar.

While some landowners awaited the advances of speculative developers and builders before entering the urban property market, others took positive steps to attract interest in their lands. Thus, strictly speaking, they went beyond the role of owners and became the promoters and developers of their own schemes. The majority of cases involved nothing more than laying out roads, drains and plots on a few fields, but at their most spectacular such positive actions led to the building of new towns from scratch. An example of the latter was the achievement of Sir George Tapps Gervis, second baronet and Lord of the Manor of Holdenhurst, who, in association with other local owners, set out to build a watering place at Bournemouth in the fourth decade of the century. It was Gervis who in 1838 commissioned a local architect, Benjamin Ferrey of Christchurch, to draw up a plan for a resort on what was then a stretch of open heathland. This initiative encouraged the adjacent owners to lay out parts of their estates. As the result of this enterprise, in the second half of the century Bournemouth grew from a settlement of less than 1,000 inhabitants in 1851 to a select resort of almost 60,000 people in 1901.

The pattern of development at Bournemouth has been outlined by Roberts (1982). From his study it can be seen that the site was divided between three main groups of owners (Figure 10.5), each of whom made a distinctive mark on the growing town. First, there were the owners of the 'enclosure estates', those who

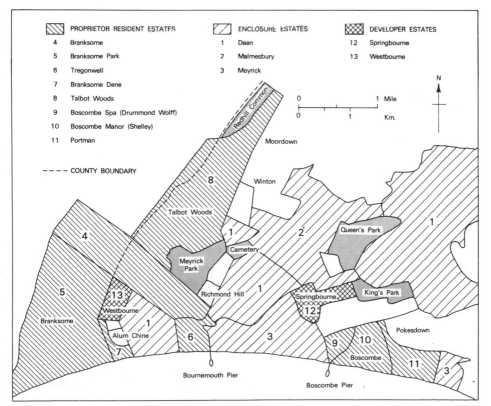

Figure 10.5 Estate ownership in Bournemouth in the nineteenth century. (after Roberts, 1982, Figure 17, 185)

were the principal beneficiaries of the enclosure of the open heathland in 1805. By this enclosure award Sir George Tapps (first baronet), the Earl of Malmesbury and William Dean received over half of the commonable land and this formed the basis of their large leasehold estates, the Meyrick Estate, the Malmesbury Estate and the Dean Estate, respectively. These owners did not live on their town estates, but each had a residence near Bournemouth. Later, some of their land was sold to other parties. In the second group were the 'proprietor resident' owners who purchased fairly large tracts of land in the fashionable town as sites for their own imposing houses with their extensive grounds. Following in the footsteps of Louis Tregonwell, who bought 40 acres at the mouth of the Bourne stream for Exeter House in 1810, were other men of means like Sir Percy Shelley, Sir Henry Drummond Wolff and Lord Portman. Finally, there were the 'developer estates' of Westbourne and Springbourne, acquired by Robert Kerley and Peter Tuck, respectively. Kerley and Tuck were building contractors who bought land for its development potential.

During the second half of the century all three types of estates were built on. Some of the enclosure lands were made available to builders from the very beginning of the town. On the Meyrick Estate, for example, the work started by Sir George Tapps Gervis in 1838 to house and cater for affluent visitors was continued by his successor Sir George Tapps Gervis Meyrick, the third baronet, and by the end of the century the development of the Meyrick land in central Bournemouth was completed. In the first instance, the proprietor resident estates were used by the owners themselves for their own houses set in secluded greenery, but in due course they too became building sites. The Tregonwell Estate pointed the way forward:

> John Tregonwell, proprietor of the Tregonwell Estate from 1846, followed the example set by the Meyrick Estate. Although the earliest lease on the Tregonwell Estate dates from 1844, development really got under way only in the 1850s and 1860s. By 1885, when John Tregonwell was succeeded by Hector Monro, the estate was fully developed, having become the first of the proprietor resident estates to be turned from a seaside retreat set in substantial grounds into urban real estate. Indeed, it had already ceased to merit the description proprietor resident estate, since the family's Bournemouth residence had been turned into the Exeter Hotel. (Roberts, 1982, 186).

As intended by Kerley and Tuck, the Westbourne and Springbourne Estates were laid out as building ground. Thus, in the course of half a century, local landowners helped to promote the building of a fashionable resort *de novo*.

Tenure

Clearly, the initial willingness of owners to release land was of importance in shaping the broad pattern and direction of urban growth, but their influence often extended well beyond that. At the most obvious and superficial it was the naming of new streets after family and friends. In Cardiff, for example, this was certainly one of the practices of the Bute estate: 'The titles of the family – Bute, Dumfries, Windsor, Mountjoy, Crichton and Mountstuart – appear on streets, places, crescents and squares. The names of their predecessors as lords of Cardiff Castle, among them Fitzhamon, Despenser, Clare and Tudor, are recalled in the streets of Riverside. In much of Cathays the streets bear the names of the manors and villages of the Bute estate in Glamorgan, while elsewhere are Sanquahar Street, Inchmarnock Street and others, named after Bute properties in Scotland' . . . and so on through the names of agents and advisers (Davies, 1981, 186–7). However, much more significant was the owner's (or developer's) influence on the form of the development through the terms of tenure granted on the building plots. A land-owner could insert clauses in leases which restricted developers and builders to a particular type and quality of houses, and thereby could determine the character of new residential neighbourhoods. In freehold, too, the owner could impose condi-tions on builders which had to be met before the freehold was assigned. Many writers have speculated on the relationship between type of tenure and quality and type of development, and although there are no firm conclusions in their studies, it is agreed that tenure was a controlling factor in some localities.

The influence of type of tenure on the quality and type of building was discussed by the Select Committee on Town Holdings which reported in the late 1880s. In its detailed examination of the principal tenures in England and Wales, it identified five main systems:

1   *Freehold*, where the fee-simple of land was sold outright to a purchaser;
2   *Fee-farm or chief rent*, where the land was granted in perpetuity, subject to a fixed annual payment;
3   *Long leasehold*, where the land was let at an annual rent for a term of 999 years or other long term exceeding 99 years;
4   *Ordinary leasehold*, by which the land was let at a rent for terms of 99 years and under;
5   *Life-lease*, where the land was let at a rent for a term of years determinable with lives, generally three (Report from the Select Committee on Town Holdings, 1889, 6).

Across the counties of England and Wales there were some regional patterns where particular types of tenure were dominant in towns: 'freehold purchase is the system adopted in by far the largest number of towns in the Northern, Midland, and Eastern Counties, although there are a few towns where the system of grants in fee-farm is in vogue, and the long-lease system is in use in many towns principally situate in Lancashire and Cheshire. There are also some notable cases where the 99 years' system prevails, such as Jarrow, Sheffield and Southport in the Northern, Birmingham and Oxford in the Midland, and Great Grimsby in the Eastern Counties. The leasehold system is far the most usual in the Metropolis and its suburbs, and in many of the towns in the South and West of England, such as Eastbourne, Folkestone, Falmouth, and Bournemouth. Life-leases are found principally in Devon and Cornwall. The towns in South Wales are almost universally built on the leasehold system, which also exists in a large number of places in North Wales' (Report from the Select Committee on Town Holdings, 1889, 9). Over and above this variety in England and Wales, the practice of feu-(fee-) farm tenure was prevalent in Scotland, which, as will be seen later, may have determined the type of buildings in Scottish towns.

Making the simple distinction between freehold and leasehold property, the Select Committee itself attempted to ascertain whether these two different tenures had a bearing on the quality of construction and the subsequent maintenance of houses. The hypothesis posed was that houses built on freehold land were not so well constructed as those upon leasehold where there were covenants relating to materials, repairs and maintenance. The evidence was inconclusive. Among that for London, for example, one witness reported that 'I have many times had to go over small freehold property, and, as a rule, I have found that worse than leaseholds', and then went on to claim that 'a proper agent finding a leasehold out of order would require it to be put to rights at once. It would be very wrong of a man to be always worrying a tenant, but directly buildings get at all out of order I have always made it a practice to have them put right' (*Report from the Select Committee*, 1887, 271). However, elsewhere in the Select Committee's reports there is evidence for London and other towns that better quality property was built on freehold land. Thus, while tenure may well have influenced individual builders in particular schemes, there was no consistent relationship across the country between the type of tenure and the quality of houses.

In similar vein, it has been speculated that the overall type of tenure may have influenced the density of housing put into building plots, with freehold leading to higher densities than leasehold. On the assumption that the small builders who bought the freeholds outright had to find more capital than their counterparts on leasehold land, it has been argued that they were obliged to make the most intensive use possible of their plots to obtain an adequate return. Mortimore, for example, has suggested that this was the case in some of the towns in the West Riding of Yorkshire, and of Bradford in particular, where high density back-to-back houses were built on freehold sites and where, in due course, this type of dwelling became the established norm for the working man (Mortimore, 1969). Conversely, in towns with leasehold tenure, as in South Wales, housing was of the less congested 'through' type with yards at the rear.

However, Daunton has shown that this argument does not hold up under close scrutiny on two counts (Daunton, 1983). In the first place, the builder did not necessarily have to pay out the full purchase price of freehold land in a lump sum. It was quite common for builders, or indeed middlemen, to pay only interest on the purchase price, or a small deposit and interest on the balance, until the house or houses had been constructed and sold, and only then would the landowner receive the sum due to him and would the land be finally conveyed. This was the procedure adopted by the Northbourne family in the disposal of freehold land in Gateshead, as shown in an example for 1875 when 'the estate agreed to sell 138 square yards of land to a builder for £69. The builder was to pay interest on the purchase price at a rate of 5 per cent per annum. By 1878, a house had been erected on the plot, which was sold for £400. The purchaser paid £331 to the builder, and £69 to the estate for the freehold. The builder had therefore needed to find only about £3 10s a year for the use of the land, rather than a lump sum' (Daunton, 1983, 66). Given deferment arrangements such as this, the developer of freehold was under no more financial pressure to build at high density than his opposite number on leasehold land.

In the second place, Daunton has dismissed the connection between freehold

Table 10.1 Tenure and housing conditions, 1911 (after Daunton, 1983, Table 4.2, 67)

	Tenure	Percentage of population in overcrowded housing
Gateshead	freehold	33.7
Plymouth	freehold	17.5
Ipswich	freehold	1.3
Leicester	freehold	1.1
Darlington	freehold	12.8
Middlesbrough	freehold	13.4
Jarrow	short lease	37.8
Liverpool	short lease	10.1
Cardiff	short lease	4.8
Sunderland	chief rent	32.6
Bristol	chief rent	4.8

tenure and small, overcrowded houses on straight empirical evidence. An examination of the type of tenure and housing conditions in selected provincial towns in England and Wales in 1911 revealed that both freehold and leasehold tenures showed highly variable percentages of overcrowding (Table 10.1). He concluded that 'tenure alone was clearly not a sufficient explanation of variation in housing conditions between English cities' (Daunton, 1983, 68).

Although type of tenure may not have dictated the density and condition of urban housing in England and Wales, there is some evidence that it did have a direct bearing in Scotland. It has been argued by Rodger and others that the 'feu' system of tenure – broadly similar to the little used fee-farm or chief rent system elsewhere in Britain – was a determinant of the distinctive type of housing built in Scottish towns, namely the high-rise tenement dwelling. In short, under the feuing system 'land was sold outright by the vendor who relinquished all title to it, but an annual payment was also made – the feu duty' (Rodger, 1983, 201), and it was this annual feu duty in perpetuity which encouraged high density development. Since the vendor had no further financial stake in the future increased value of his land or of the buildings upon it, it was in his long-term interest to negotiate as high an annual feu as possible from the purchaser at the time of sale. By building high-density tenements, the purchaser, or others to whom the land passed by the process of sub-feuing, could recoup the high feu duty from many tenants and make a profit. This was the explanation given for the widespread construction of tenements by the Scottish Land Enquiry Committee in its report of 1914:

> When the ground rent exceeds a certain figure (£20 to £25 an acre per annum), the erection of cottages at reasonable rents becomes impossible and it is necessary for the builder to erect tenements in order to spread the high cost of the land over a larger number of houses. The exaction of high feu duties by persons granting feus on the outskirts of Scottish towns has thus a tendency to necessitate the erection of tenements in order to make it profitable to develop the land subject to these payments. The high price of land and the erection of tenement housing react upon one another, that is to say, the high price of land requires the erection of tenements in order to make the maximum use of it and to spread the burden of the feu duty over as many payers as possible, and conversely the power to erect tenements and to impose upon the land a very considerable property maintains the high value of the land. (Scottish Land Enquiry Committee, 1914, 293)

Thus, while tenements were not exclusive to Scotland, their concentration in large numbers there and the evidence contained in contemporary descriptions and in official enquiries into land tenure and house-building, indicate that feuing was at least a part determinant of high-density development.

While there is debate over the precise relationship between the type of tenure and the type, quality and density of property at the broad scale, it is quite clear that individual owners did employ restrictive clauses in the assignment of land to try to control the quality of what was built on it. On Lord Calthorpe's Edgbaston estate in west Birmingham, for example, the aim of the agent was to create an exclusive suburb in the early part of the century, and some success was achieved in this by the use of covenants in the leases (Cannadine, 1980). One of Cannadine's illustrations, a lease granted in 1828 with a restrictive covenant containing a list of undesirable properties, clearly shows the prevailing attitude: 'Any small dwelling house or houses of the description of labourers' or poor persons' houses or which shall be occupied by labourers or poor persons, nor any other small dwelling house or houses, nor any workshop or workshops or other kind of shop or shops, nor any place or places for carrying on any trade or manufacture, nor any brewshop, ale house, tea garden, public strawberry garden, or any other place of public resort or

Table 10.2 Value of houses built on the Edgbaston estate, Birmingham, 1786–1914 (after Cannadine, 1980, Table 7, 113)

Value of house (£)	No. of leases granted	%	No. of houses authorized	%
1,500+	165	12.85	189	6.65
1,000–1,499	180	14.02	247	8.69
500–999	468	36.45	944	33.23
0–499	471	36.68	1,461	51.43
TOTAL	1,284	100.00	2,841	100.00

amusement whatsoever . . . nor any other erection or building whatsoever which shall or may be deemed a nuisance or otherwise injurious to the said Lord Calthorpe' (Cannadine, 1980, 119). During the century, the estate was neither able to develop all of its land nor attract only the most expensive properties, as can be seen from Table 10.2, which shows that half of the houses were for the lower middle and prosperous working classes (houses valued at less than £500), but by preventing the intrusion of industry, the Calthorpes still housed relatively large numbers of the aristocratic and upper middle classes and were able to preserve the high social tone of Edgbaston (Figure 10.6).

The practice of owners seeking to maintain the select character of houses through the inclusion of restrictive covenants in leases was widespread in exclusive neighbourhoods. Returning to the example of Bournemouth again, in the last quarter of the century restrictions were applied to a large part of the town. Villas built on the Meyrick, Dean, Tregonwell, Boscombe Spa, Branksome and Branksome Park

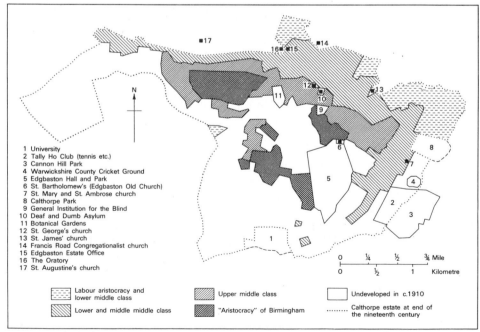

1 University
2 Tally Ho Club (tennis etc.)
3 Cannon Hill Park
4 Warwickshire County Cricket Ground
5 Edgbaston Hall and Park
6 St. Bartholomew's (Edgbaston Old Church)
7 St. Mary and St. Ambrose church
8 Calthorpe Park
9 General Institution for the Blind
10 Deaf and Dumb Asylum
11 Botanical Gardens
12 St. George's church
13 St. James' church
14 Francis Road Congregationalist church
15 Edgbaston Estate Office
16 The Oratory
17 St. Augustine's church

Labour aristocracy and lower middle class
Lower and middle middle class
Upper middle class
"Aristocracy" of Birmingham
Undeveloped in c.1910
Calthorpe estate at end of the nineteenth century

Figure 10.6 Residential zones in Edgbaston, Birmingham, *c.* 1910. (after Cannadine, 1980, Figure 12, 114)

estates (see Figure 10.5) were strictly supervised by the estate offices: 'Their building leases specified the minimum value of the house to be erected, the standards of construction and sanitation to be met, the achitectural style, and the time within which the building was to be completed. Restrictive covenants in the leases granted to tenants imposed a wide range of duties and prohibitions. Take, for example, the lease granted to S. J. Stevens by the Dean Estate on 25 March 1876. Stevens was prohibited from making additions or alterations, from touching the trees in the garden, from using the building for any commercial purpose or as a school, or a chapel, or for holding public sales or auctions. He was obliged to paint the outside every third and seventh year, to keep the house in good structural order and to make any repairs requested by the estate within three months, to insure it at an office specified by the estate and to contribute to the upkeep of parks and pleasure grounds on the estate. The estate reserved a right of entry for inspection twice a year and access to trees at all times. All fees and costs were to be born by the lessee, and the estate had a right of re-entry upon 21 days default on the payment of ground rent.' (Roberts, 1982, 189) The adoption of similar leases in the villa districts helped to preserve the resort's reputation as a select watering place.

However, strict control over the quality and type of property was not confined to leasehold. The owners of land that was to eventually sell as freehold were equally anxious to preserve the value of their plots in prestigious neighbourhoods. This was achieved by disposing of their land to builders as leasehold in the first instance, but giving them the option of purchasing the freehold after a certain number of years. In the initial lease the owner would lay down his constraints on the builder. This was the procedure followed in the development of part of another south coast resort, Brighton. For example, on the Stanford estate on the western edge of the growing town in the last quarter of the century 'all deeds had the standard specification which obliged the builder to lease for 99 years at an agreed rent but with the opportunity to purchase within seven years at a purchase price which was the equivalent of 25 years rent. The builder had to complete the shell of the property within seven years to an agreed minimum value and observe the building line prescribed for that road by the estate. The conveyance by the Stanford Estate was to the purchaser of the completed house, which could be the builder but was often the person to whom he had sold it' (Farrant *et al.*, 1981, 38).

Restrictions were not confined to the most exclusive neighbourhoods, but were often used in working-class districts to prevent the building of very poor quality, high-density dwellings. This can be seen on the Ramsden Estate in Huddersfield in the second half of the century (Springett, 1982). By 1850 most of the land in and around the town was in the hands of four estates, the Ramsden, Lockwood Proprietors, Thornhill and Kaye, with the Ramsden family dominating the town centre (Figure 10.7). While the Ramsden estate was quite prepared to let land to builders on a variety of terms, including 60-year renewable leases, tenancy at will, and 99-year and 999-year leases, it was also concerned to secure the erection of good class buildings on its property, and thus it gradually introduced restrictive covenants into its leases. By the 1850s it had become established policy to prohibit the erection of the cheaper back-to-back houses favoured by builders. This resistance to the market had three effects on development. First, it encouraged building for the lower middle classes on Ramsden land, with proportionately higher rents. Second, poorer working-class housing, and particularly back-to-back, was deflected to other estates, often at some distance from the town centre. Third, the intensity of development was comparatively low and stood out amongst that which was taking place on similar sites on adjacent estates. One example of this contrast was the Primrose Hill area just south of the confluence of the rivers Holme

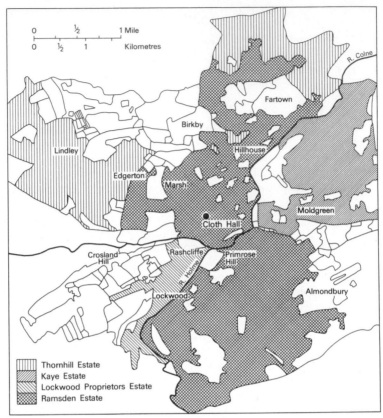

Figure 10.7 The pattern of landownership in Huddersfield in 1850.
(after Springett, 1972, Figure 1, 130)

and Colne where Ramsden property along Whitehead Lane abutted the Lockwood Proprietors estate. Following the bridging of the Holme in 1861, when the area came under particular pressure to house workers from the now accessible Rashcliffe mills, the two owners showed different responses which produced clear contrasts on the ground. 'Almost immediately land was leased from the Lockwood Proprietors and between 1861 and 1865 an inn and 76 back-to-back houses were built. Only six through houses and one shop were built on the Ramsden portion of the road. On the Lockwood Estate the average area per house was 65 square yards; on the Ramsden Estate 253 square yards. On the former estate the rent charged was only 1.65d per square yard compared with 1.75–2.25d (2.75d after 1867) on the Ramsden Estate.' (Springett, 1982, 139) Eventually, the Ramsdens capitulated on the demand for cheaper back-to-backs, but not until 1900 and not before the clauses in their leases had had an impact on the form and intensity of development in central Huddersfield.

Land holdings

It is evident from the examples given above that the landowner was one of the primary decision-makers shaping the spread of towns, and the type and quality of property built on newly released building plots. In similar manner, developers determined the character of neighbourhoods by the restrictions they imposed on builders. Before turning to look at the builders themselves, there is one further matter to be explored under the general heading of landownership, the influence of the shapes and sizes of the pre-existing land holdings on the eventual arrangement of streets and houses.

In some cases, of course, the underlying pattern of fields and plots was of no real consequence to the eventual urban layout. In the extreme, the pre-existing field boundaries could be almost completely obliterated by a planned arrangement of streets and dwellings. Such a drastic impact was achieved where the land was in the hands of the single landowner or developer who had the vision of building an entirely new town or suburb, and who could work to a surveyed plan in an orderly and systematic manner. In these circumstances it was common for a grid-iron layout of streets and measured plots to be superimposed on the site. This was the

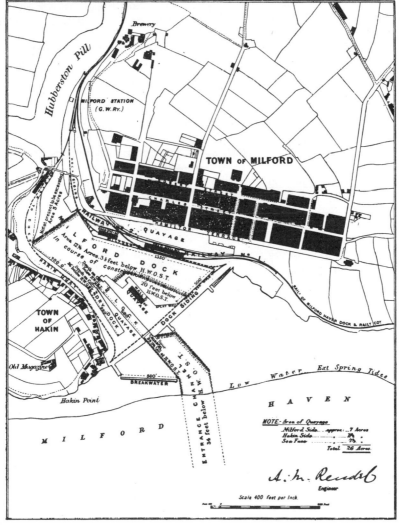

Figure 10.8 Generalized map of the planned town of Milford Haven, Pembrokeshire, *c.* 1885.
(after Rees, 1957, Figure 5, 82)

sort of development experienced by those places which have been described as 'planned' towns. One example is Middlesbrough, where Joseph Pease and his partners in the Owners of the Middlesbrough Estate Company bought 500 acres in 1829 and drew up plans for a coal port on what was agricultural land, mainly salt marshes. On the driest site 'Pease planned a gridiron town of 32 acres. He divided his neatly-drawn square into 125 plots, each 200 feet by 60 feet, centred on a core of church and market, and with a spare section in the north-east for a burial ground. The wide central square, on the peak of the knoll, was the intersection for four main streets, each 56 feet wide; between each of these there ran subsidiary streets in parallel ranks, 36 feet wide' (Bell, 1972, 179). The same story, in whole or in part, was repeated in other places (Figure 10.8). In short, the pre-existing cadastral patterns were swept aside where strong control was exercised through development plans.

More generally the potential building land in and around towns was subdivided into small parcels and held by innumerable small owners. Development of this land proceeded in a piecemeal manner, both spatially and temporally, as and when individual plots were released to builders. This was the case in Bradford, where before 1900 few large estates were available for building and all types of development had to make use of smallholdings (Mortimore, 1969). The large estates round the town, such as the Bowling Iron Company, the Low Moor Company and Powell, were unwilling to release their land and thus stood out as gaps in the continuity of the built-up area. Some of these have survived to the present as public open spaces or as green belt. However, the smallholder was the most common landowner in this part of the West Riding in the middle of the century, with some 85 per cent of the landholdings of 6 ha. or less in size. Particularly distinctive were the long, narrow crofts extending behind houses which had been used for part-time agricultural activities and tentergarths in the domestic weaving economy of the smallholders. As observed by Mortimore, the town advanced across these parcels of freehold land: 'with the disappearance of the smallholders following industrialization, these crofts became available for building and were characteristically developed in the form known as yard property, having houses along one or both sides, backing on to similar terraces in the next yard and often reached from the street by means of a covered arch' (Mortimore, 1969, 109). Thus, the rather fragmented land ownership pattern was carried forward in the townscape as small builders covered each plot with high density, mainly back-to-back, houses.

In his study of back-to-back housing in Leeds, Beresford has also drawn attention to the influence of the fragmented ownership pattern on residential development (Beresford, 1971). The enclosure of the open fields around the town in the later Middle Ages had produced the same long, rather narrow smallholdings, and it was these separate fields which earliest came on to the market. The typical building plot available after 1750 was from 120 to 200 ft wide, and about 600 ft long; the average size of the 54 fields designated as 'building ground' in the Tithe Award of 1847 was just over one and a quarter acres. In the construction of roads and in the alignment of houses builders adhered to the boundaries of these elongated plots and thus development proceeded in a rather piecemeal and disjointed manner. With regard to the layout of streets, it was not uncommon 'to find their ends blocked by a wall, by a change in level, or by the sides of other houses in another unconnected street. The line of blockage is simply the old field boundary. On one side a developer has set out his streets: and on the other side a different developer at the different time has set out his; and never the twain should meet' (Beresford, 1971, 108). The shapes of the building plots also produced some curious types and shapes of houses as developers tried to squeeze in the maximum

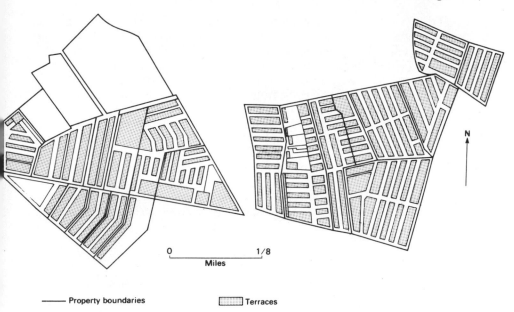

N
↑

0 —————— 1/8
Miles

———— Property boundaries   [▨] Terraces

Figure 10.9 Effects of land holdings on the alignment of terraces in two areas at Armley, Leeds.
(after Ward, 1962, Figure 4, 155)

number of dwellings. For example, where there was insufficient width to build a line of back-to-backs at the edge of a plot, the 'half-back' was constructed, that is, the builder 'put up half a back-to-back as far as the peak of the roof and then dropped vertically down what would otherwise have been the partition wall between back-to-backs' (Beresford, 1971, 108).

The fragmented ownership of land continued to thwart the orderly development of Leeds as it pushed southwards in the second half of the century. Before residential building took place in the three southern townships of Hunslet, Holbeck and Armley, over 80 per cent of their holdings were less than 10 acres in size (Ward, 1962, 152). As shown by Ward (Figure 10.9), the boundaries of these parcels determined the arrangement of streets and buildings: 'Each variation in the alignment or direction of terraces and streets coincides with the pre-existing property lines . . . The terraces are successively reduced or increased in length with each irregularity in the property line and are often reduced in length to only one or two houses' (Ward, 1962, 152). In passing it can be noted that this type of development characterized the fringe of Nottingham when it eventually broke out of its medieval straitjacket. On the allotments set out by the enclosure commissioners on the former common lands '. . . each one of the new owners had an unfettered right to develop his allotment subject to the provision of the Enclosure Act. Since there were about four hundred owners there were, in effect, four hundred little town planners at work, each busily engaged in the development of his allotment according to his own ideas and without reference to what was taking place on the allotment of his neighbour' (Chambers, 1952, 13).

The smallholdings of Bradford and Leeds had their even smaller counterparts in the burgage plots, inn yards, gardens and orchards around the cores of so many of Britain's towns. Indeed, the smallest crofts which flanked the central streets of Bradford and Leeds were themselves nothing more than backyards. It was often the case that these vacant intra-urban plots were the first to be seized upon by builders in the early stages of industrialization. Their very smallness and sometimes higgledy-piggledy shapes gave rise to quite distinctive residential cells or 'courts' tucked away behind the main thoroughfares, but, in general, they were long and narrow, and entered by a tunnel or alley from the street. This process of intra-mural infill characterized towns of all sizes, from small but growing market places, right through to the major industrial centres.

At the level of the small country town, Conzen's study of Alnwick in Northumberland, a town with a population of about 6,500 at mid century, has shown how this process of infill – what he called 'repletion' – operated (Conzen, 1960). One of his illustrations was a burgage off Fenkle Street which had been owned by George Selby, a well-known and respected attorney. In the late eighteenth century this enclosed plot contained a dwelling house and yard, and a long 'garth' or garden at the rear; less than 15 per cent of the total area was covered by buildings (Figure

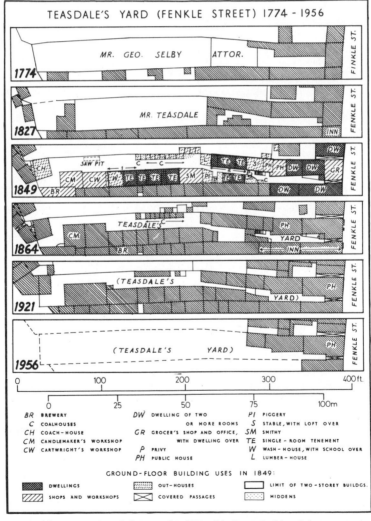

Figure 10.10 Repletion in Teasdale's yard (off Fenkle Street), Alnwick, 1774–1956. (after Conzen, 1960, Figure 14, 68)

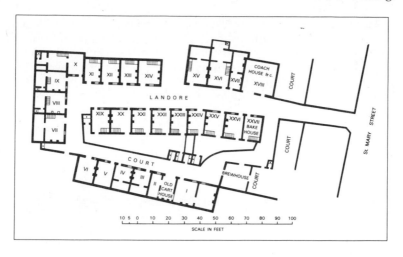

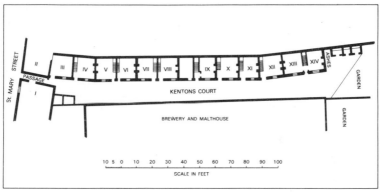

Figure 10.11 Landore court and Kenton's court, Cardiff. (after Rammell, 1850, Plans 4 and 5)

10.10). By 1827 its ownership had passed to a Mr Teasdale, and the building coverage had increased to about 35 per cent through the construction of small back-to-back houses. Repletion continued with the erection of more back-to-backs and small workshops so that by 1849 the building coverage was 62.9 per cent. Although the amount of vacant land had been reduced to a very narrow alley and yard, reached through a covered passage from the street, the outline of the original burgage was clearly preserved in the alignment of the new premises. In this manner, even the smallest parcels of land were put to use.

It is clear from contemporary descriptions and plans that the processes identified by Conzen in the market town of Alnwick were intensified on the small plots in the kernels of the growing manufacturing towns and ports where the demand for cheap housing was much greater. The health reports for mid century are full of examples of infilling and overcrowding, where builders squeezed in high-density courts on vacant yards and gardens. In each case, the shape of the building plot dictated the shape of the court, as can be seen in Figure 10.11 which shows two of the courts leading off St Mary's Street in Cardiff in 1850. In his health report of that year, Rammell described the layout and condition of both. Of Landore court, he said it 'is built upon an irregular plot of ground, measuring about 160 feet long by 120 wide. It is closely built up on three sides; and, in order to make the most of space, an additional or middle row runs down the centre, between the two side rows. The passage or footway through the court is about 15 feet broad, in some places as narrow as 10 feet . . . This court leads out of St Mary's Street by a narrow passage, and is not a thoroughfare. It contains 27 houses, generally of two rooms each' (Rammell, 1850, 39). Kenton's Court on the other side of St Mary's Street had barely enough room for a row of houses: 'It is a long narrow passage, entered by an archway from the street, between 4 and 5 feet wide. The plot of ground occupied by the court is about 200 feet in length, by 27 and a half in width, of which 14 feet is occupied by the passage or footway in front of the houses. This court is very much confined by the dead wall of the brewery and malthouse on the side opposite to the houses. There is no thoroughfare through the court' (Rammell, 1850, 40).

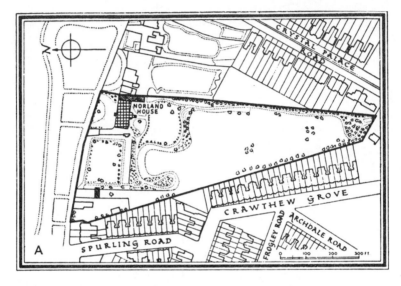

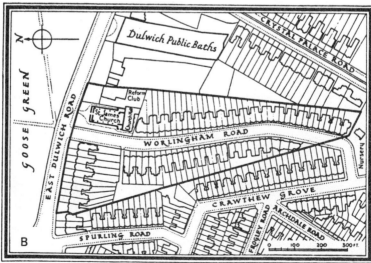

Figure 10.12 Repletion in Camberwell: **A** Worlingham Road estate in 1880; **B** Worlingham Road estate in 1895.
(after Dyos, 1961, Figures 7 and 8, 108)

While repletion was particularly characteristic of intra-mural gardens and yards in the early stages of urban growth, the same process was in evidence later when the grounds belonging to detached mansions and villas in the inner suburbs were taken over by builders. By this process the gardens and ornamental shrubberies of one phase of suburban residential development became the building plots of a later phase of infilling as the demand for suburban residences increased. Again, the shapes of the plots dictated the layout of the new terraces and cul-de-sacs, but sometimes the original villas themselves were demolished. This pattern of second-ary development is well illustrated in Dyos's study of internal residential change in Camberwell, what he described as 'building estates on back gardens' to accommo-date 'the advancing cliffs of the new suburbia' in the late nineteenth century (Dyos, 1961, 105–6). One of the examples for East Dulwich is reproduced in Figure 10.12. Here an estate of about eight acres had been bought by Thomas Baily, farmer and stockholder in the East India Company, and partially developed with large houses between 1804 and 1837. By the 1870s the estate had been virtually encircled by the advancing tide of much more modest semi-detached and terraced villas, which 'naturally put a premium upon the serpentine walks and shrubberies of the few houses whose grounds covered the bulk of the original estate, and the sale of the largest of them, Norland House, in 1877 was a preliminary to more intensive use of the ground' (Dyos, 1961, 107). Three years later the property was auctioned again, this time as a freehold building site, with a suggested layout for new houses. In due course 'a terraced street, named Worlingham Road, swept through these triangular-shaped grounds along the suggested route between 1881 and 1887 and Norland House itself, dismembered of its grounds, survived the operation to house from 1881 a small congregation of the Church of Scotland' (Dyos, 1961, 107). Thus in suburban districts too, the process of repletion was constrained by the sizes and shapes of former gardens. In short, the residential layout imposed by the first generation of builders was re-worked at higher density by a second generation.

Conclusion

This chapter has looked at the role of one of the key decision-makers, the landowner, in the conversion of land to urban use. By referring to selected examples, it has been shown that owners (or, for that matter, their agents and developers) controlled the direction and rate of urban encroachment, and through the use of covenants in the tenures granted to builders, dictated the type, quality and density of property built on their plots. While these influences had a direct impact on the general layout and character of towns (or at least their various neighbourhoods), landownership was also often carried forward into the physical form in a very precise manner in that the shapes of the underlying plots released to builders determined the subsequent arrangement of streets and houses on the ground. As has been shown, in those places where small patches were held by numerous owners, the inherited cadastral pattern led to unco-ordinated urban spread when builders implemented their own schemes with little or no regard to what was occurring on adjacent plots. Having looked at these influences, it is proposed to examine the activities of the second major figure in Springett's decision-making framework, the builder. This will be done in the next chapter.

11

The supply of the urban fabric

Given the availability of building land, fettered or unfettered by restrictive clauses and covenants, its development involved the putting to work of capital and skills with the eventual aim of financial gain. As Tarbuck's *Handbook of House Property* put it

> The richest crop for any field;
> Is a crop of bricks for it to yield;
> The richest crop that it can grow;
> Is a crop of houses in a row.
> (Quoted in Halstead, 1982, 57)

However, returning to the point made in the introduction to the previous chapter, the transformation of a greenfield site into a completed street of houses was by no means a standard procedure whereby the key decision makers went about their affairs as separate individuals in a well organized sequence. On the one hand, the landowner himself could finance and oversee an entire development, paying for the appropriate professional and craft skills as and when required, and to that extent could exercise a degree of central control and organization. Quite legitimately, he could be described as owner, developer, financier and builder. On the other hand, the development of a site might be sponsored by numerous individuals – developers, employers, solicitors, architects, small investors, building societies, tradesmen – and subject to separate decisions and changing personal fortunes and commitments. Thus, there were no hard and fast lines between the main decision makers. That there were certain key roles to be performed is not in doubt, as indicated by Springett in Figure 10.1, but there was a great deal of overlap between the roles played by individuals. As the century progressed, the functions of some figures became much more clearly defined and specialized, particularly in the technical professions engaged in the building industry, but the distinctions between owner, developer, financier and builder often remained blurred, something noted by Dyos in his study of speculative suburban building around London: 'Here were innumerable situations in which the functions that were involved in developing building estates, and which are capable of being identified in isolation, actually overlapped. It was not the custom of landowners, for example, to stick merely at supplying land against ground rents; developers did not stop short when they had pegged out building plots; the suppliers of finance were not always prepared to keep strictly to their financial operations; nor were the business operations needed to open up an area by transport developments kept strictly separate from the building spree. In fact, situations became very quickly confused, the more so because the creation of titles to wealth in new districts – improved ground rents as well as commercial property – were passed rapidly from hand to hand' (Dyos, 1968, 645–6). In its concern with this process of development, this chapter is arranged around the second key figure in Springett's decision-making framework, the builder, and those who provided support services for him, but, as stressed above, there was a blurring of roles, and it will be necessary to return to this point in the discussion. While this chapter again concentrates on the housing sector, it should be noted in passing that the nineteenth century was an era for both grand public buildings – town halls, market halls, hospitals, churches – and impressive commercial and manufacturing premises. Many of the functions and activities discussed in Chapters 6, 7 and 8 were accommodated in massive Victorian edifices which still dominate today's townscape, but the building of these structures will not be explored here.

Following the release of blocks of land for building purposes, but before building work commenced, a developer may have intervened as an intermediary or middle-man between owner and builder, deciding on the potential of a site and then getting it ready for builders. He would be committed to some financial outlay for the purchase or lease of the site itself and for the installation of basic facilities, such as roads and pavements, drains and perhaps water supply. A land surveyor might also be engaged to measure and mark out individual building plots. At that stage he would start to recoup his capital and make a profit by selling or sub-letting parcels to builders. Of course, many landowners themselves assumed the role of develop-ers. The discussion in Chapter 10 showed that some owners took a direct interest in the layout of their sites and in the type of property built, and it was common for the agent of the large owner to act as the developer. However, where owners played a much less active part, developers came into play in their own right, on both freehold and leasehold land.

Like the landowner, the developer too often wore several hats and went well beyond the stage of site preparation into building operations. One example documented by Dyos (1968) was Charles Henry Blake, a retired Indian civil servant, who was primarily a developer in the Notting Hill district of West London in the 1850s and 1860s but who also acted as owner, financier and builder in his dealings. Under the initial stimulus of his solicitor, he put his own and borrowed money into land, he financed builders, and he hired builders on his own account. Typical of the developer's enterprise was his promotion of the City and Hammer-smith Railway (1864). Using borrowed money, he bought land ahead of the railway and then proceeded to sell and lease plots for building at a profit. Dyos has calculated the balance sheet of his transactions thus: 'Blake's total investment (plus interest payments) in the Portobello estate, as it was known between 1863 and 1879, was about £95,000, laid out on nearly 60 acres; and in return he got £56,000 from sales of land, £54,000 from sales of finished houses, and a little over £2,000 from rents. He still held some of the land, which was then worth three times what he had paid for it' (Dyos, 1968, 647). By the time of his death in 1872 he had a personal fortune of £35,000 and a real estate of £473,000. This one case well illustrates the blurring of roles of individuals referred to earlier.

Builders

The task of erecting property on the building plots set out by developers was the function of the builder, but he (or she in a minority of cases) was not necessarily a craftsman in the building trades. While those with building skills were involved, many 'builders' were relatively small-scale speculators from a wide variety of backgrounds who used their own or borrowed money to purchase materials and hire labour to build houses, shops and workshops and who then sold or let them to others, or (but not generally) occupied them themselves. The evidence for Huddersfield in Springett's study (1979) shows that members of the building trade itself were not the main speculators; rather, merchants, professionals, shopkeepers and general tradesmen masterminded operations and employed craftsmen to do the building work. However, as the century progressed, those in the building trades played a more prominent part, seeing the whole process through, from the initial acquisition of land to the completion of houses. Whatever the callings of those described as builders, two points to stress about them are that they were predominantly speculators and that individually they made small contributions to the housing stock. Daunton, for example, noted in his analysis of Cardiff that they were 'pre-eminently speculative builders' composed of 'small-scale units' (Daunton, 1977, 89), and Rodger observed that at the end of the century 'housebuilding in Victorian and Edwardian Scotland was biased heavily towards small operations. Just over half, in fact 52.9 per cent, of housebuilding in 102 burghs was undertaken as "one-off" projects' (Rodger, 1979, 227). Some of Rodger's data for Scottish burghs are reproduced in Table 11.1. The generally small size of the majority of building operations has been substantiated by the evidence compiled for a large number of cities in Britain.

One of the most comprehensive statements on the size and activities of speculative builders is Dyos's classic study of the growth of Camberwell (Dyos, 1961). For the peak years of building activity in the suburb, 1878–80, he used the District Surveyors' monthly returns to calculate the number of active firms and the number of houses under construction (Table 11.2). Of the 416 builders of 5,670 houses, over half built no more than six houses and nearly three-quarters of them built no more than 12 in the three years. Many streets were put together in a piecemeal, unco-ordinated manner as numerous builders took up their own plots almost at random; further 'there were also a good number of roads in which up to 40 or 50 years elapsed between the filling of the first and the last building plots: on these the builders' names were practically legion' (Dyos, 1961, 126). Thirty years earlier, the small builder had been even more prominent. Dyos's evidence for 1850–52 shows that over 90 per cent of builders were engaged in building 12 or fewer houses and none built more than 30 houses. Dyos also showed that the majority were local men: 'The great majority of builders had local addresses, and very few whose headquarters were not in Camberwell came farther than Kennington, Brixton, Walworth, Bermondsey, or New Cross: those that did come from farther afield

Table 11.1 The scale of building projects; Scottish burghs, 1873–1914 (R. G. Rodger, 1979, Table 1, 227)

| | | Proportion of houses in house-building project (per cent) | | |
		1 house	2 or more houses	3 or more houses
Cities	(3)	39.58	60.42	21.11
Major burghs	(30)	53.87	46.13	20.36
Small burghs	(69)	74.83	25.17	3.60
Scottish burghs	(102)	52.91	47.09	17.19

Table 11.2 House-builders in Camberwell, 1878–80 (H. J. Dyos, 1961, Table 4, 125)

Size of business (number of houses under construction)	Number of firms	%	Number of houses	%
1–6	220	52.9	699	12.3
7–12	82	19.7	769	13.5
13–18	38	9.2	594	10.5
19–24	29	7.0	620	10.9
25–30	11	2.6	304	5.4
31–36	5	1.2	162	2.9
37–42	5	1.2	199	3.5
43–48	6	1.4	268	4.7
49–54	2	0.5	105	1.9
55–60	3	0.7	169	3.0
over 60	15	3.6	1,781	31.4
TOTALS	416	100.0	5,670	100.0

were generally building on contract and not on speculation. Probably nine out of ten of all the builders of houses in Camberwell in the 1870s were in fact building on speculation, and practically all the remainder were building under contract for some other builder or small capitalist whose venture was usually speculative' (Dyos, 1961, 125). In the development of east London, too, it has been shown that the majority of builders were local men who speculated on a small number of houses. On the 90 acres of land in Stepney belonging to the Mercers Company the average commitment of the 78 builders between 1817 and 1850 was less than six houses, and on the Cotton Estate in the 1870s, 295 plots, or 55 per cent, were leased for the building of only one house (Halstead, 1982, 54).

Further evidence on the structure of the building industry has been compiled by Aspinall (1978). He examined the fortnightly minutes of the New Works and Plans Sub-committee of the Borough of Sheffield for the period 1865–1900. This body, which came into being in October 1864 to administer the first building by-laws, scrutinized applications for new property and new or extended streets. From the minutes for the 36 years he identified 4,069 separate owners of house plans whom he thought were made up of three broad groups of people. The first group comprised the speculative builders who themselves built houses for sale or rent. Secondly, there were the speculative developers, those who assembled resources and craftsmen to build property which was then sold or rented. 'They included, for example, Thomas Steade, owner of plans for 309 houses and 25 streets; J. Hayhurst, owner of plans for 83 houses and 25 streets; F. U. Laycock, owner of plans for 103 houses and 15 streets; and perhaps many of the 17 other persons who owned plans for 16 or more houses and four or more streets. Then there were the landowners who took on at least some part of the role of developers, notably, the Duke of Norfolk, owner of plans for 83 streets; M. G. Burgoyne, owner of plans for 52 streets; and Earl Fitzwilliam, owner of plans for 28 streets' (Aspinall, 1978, 34). Finally, many of the owners of plans of one or a few houses may well have been intending owner-occupiers or very small-scale speculators. However, it must be stressed that this was an arbitrary grouping based on impression rather than hard fact since few details about the owners were recorded in the minutes.

Aspinall used the numbers of plans held by these various owners as his measure of the size of building concerns (see Table 11.3). What is immediately apparent is the dominance of those building small numbers of houses. More than half of the

Table 11.3 House-builders in Sheffield, 1865–1900 (P. J. Aspinall, 1978, Table 2, 41)

Size of business (No. of houses in plans)	Number of firms Absolute frequency	Relative frequency (%)	Cumulative frequency (%)
1	942	23.2	23.2
2	765	18.8	42.0
3	336	8.3	50.2
4	402	9.9	60.1
5	176	4.3	64.4
6	183	4.5	68.9
7	103	2.5	71.4
8	147	3.6	75.1
9	76	1.9	76.9
10	76	1.9	78.8
11	46	1.1	79.9
12	67	1.6	81.6
13	42	1.0	82.6
14	43	1.1	83.7
15	32	0.8	84.4
16	39	1.0	85.4
17	21	0.5	85.9
18	26	0.6	86.6
19	18	0.4	87.0
20–29	175	4.3	91.3
30–39	76	1.9	93.2
40–49	54	1.3	94.5
50–59	39	1.0	95.5
60–69	31	0.8	96.2
70–79	25	0.6	96.8
80–89	17	0.4	97.2
90–99	14	0.3	97.6
100+	98	2.4	100.0
TOTALS	4,069	100.0	100.0

builders put up no more than three houses, and three-quarters no more than eight houses. His is further evidence of the generally small size of building operations. Later in the century there was an overall shift towards larger concerns, perhaps in response to the expanding demand for suburban houses. Figure 11.1 shows that those firms building less than eight houses generally declined throughout the period, while larger firms increased in number. Nonetheless, even at the end of the century small firms building less than eight houses were still in the majority, and those building three houses or less made up a significant proportion of the total number of firms.

While small builders were most numerous, the evidence of Aspinall, Dyos and others indicates that a small number of large firms were active and that as the century wore on their share of the market increased. One striking feature of Table 11.2 is that 15 firms (only 3.6 per cent of the total) were responsible for 1781 or 31.4 per cent of the houses. In his review of empirical studies, Doughty concluded that 'the output of the larger firms exceeded that of the smaller, despite the latters' numerical superiority. Moreover, this imbalance became more pronounced as the century progressed, for the numerical domination of the small builder was gradually eroded, giving further impetus to the domination of output by larger operators' (Doughty, 1986, 7). Without doubt, some builders were able to sustain successful local businesses over fairly long periods, sometimes over generations, and completed whole streets and even districts. In Camberwell, the three largest builders identified by Dyos, J. Dadd of Cemetery Road, Cooper and Kendall of Queen's Road, and W. Stubbs of Lambeth, built 349, 296 and 243 houses respectively during the 1870s decade. Their success came from spreading their interests fairly widely within the locality and from building substantial numbers of properties on particular estates with which their names became associated. Stubbs, for instance, was very active on the Rosemary Branch and the Vestry Road estates. Thus, as indicated in Table 11.2, a small number of resilient firms accounted for a high proportion of the total housing stock.

These larger firms which emerged out of the mass of small speculators retained their own more or less permanent body of labourers and craftsmen and undertook

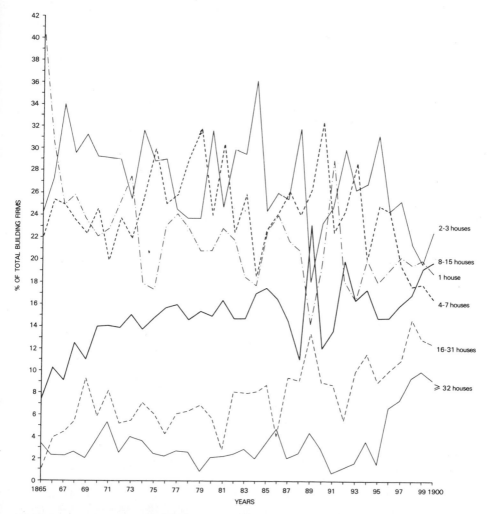

Figure 11.1 Percentage of building firms in different size groups, 1865–1900. (after Aspinall, 1978, Figure 9, 42)

residential schemes and major public works. The most successful have been described as 'master builders'. Although there is some evidence that they operated in the eighteenth century, in the building of military barracks for instance, it seems that most businesses were established during the nineteenth century in response to the steadily rising demand for all kinds of buildings. The large sums of money made available for new private and public structures, such as hospitals, colleges, schools, warehouses, docks and bridges, encouraged the growth of integrated businesses which could tender for entire projects.

At the pinnacle of the industry was the small band of master builders who were at work in London, where they were able to take advantage of the substantial demand for town houses from the well-to-do and to profit from the steady stream of 'improvements' and institutions which were needed in the growing metropolis. Although few in number, they left an indelible mark on the capital city. The prime example was Thomas Cubitt, an archetypal entrepreneur who rose to fame and riches from humble beginnings (Hobhouse, 1971). Born in rural Norfolk in 1788, he trained as a carpenter in London, perhaps working alongside his father. On the death of his father in 1806 he signed on as a ship's joiner for a voyage to India, which enabled him to save enough money to start on his own as a carpenter at Holborn when he returned to the capital. Initially he practised his own trade but soon took on the title of 'builder', at first employing sub-contractors in other trades, and then his own workforce of craftsmen and labourers. He undertook speculative projects and contracts, housing and public works. So successful were his early ventures that by 1820 he had the financial support, experience and workforce to embark on some of the most prestigious residential developments in London. From his new purpose-built workshops in Gray's Inn Road, he commenced work on the Bedford estate, and then on parts of the Grosvenor and Lowndes estates in Belgravia and the Neathouse estate of Lord Westminster in southern Pimlico. By the middle of the 1820s Cubitt employed at least 700 men, and catered for all aspects of the building trade, even supplying the shrubs and plants in the newly completed gardens and squares from his own nurseries.

While master builders were distinguished by the large sizes of their building projects and skilled workforces, and by their involvement in public and private contracts, in their day-to-day work in the field of speculative development they had the same basic requirements, albeit on a larger scale, as the small speculators. On the one hand, there was the same need to negotiate with landowners to acquire sites on acceptable terms; on the other hand, there was the same need to obtain financial backing. These can be illustrated by returning to look at the business of Thomas Cubitt. Prior to building, he tried to negotiate for sites on terms which would allow him time to complete the houses before the full rent fell due. This is well shown in one of his early approaches to the Bedford estate (which, incidentally, was initially refused) for land in Bloomsbury in which he asked for a shifting rent:

> Gray's Inn Road
> 17th May 1820
>
> Gentn.
>   I beg to propose to you the following terms for taking the Ground on 99 years Lease and building on the south side of Tavistock Square; to have 100 feet in Depth from front of Houses and 220 feet in Frontage and to build ten houses thereon agreeable to your Plan; the whole to be covered in by Midsummer 1821 and five to be finished by the following Christmas, the others within twelve months following; and to make at my Expence, the 220 feet of Sewer in Front. To have the three first years at a Pepper Corn Rent, the three next years at a Rent of £100 per ann^m and the seventh year and so on for the Remainder of the term one hundred and eighty pound per Annum.

I also engage to build the Houses in a respectable and substantial manner to the satisfaction of Mr. Stead [the Estate Surveyor].

I have the honor to be
Gentlemen
Your obed[n] Servant
Thos. Cubitt
(cited in Hobhouse, 1971, 62)

For schemes such as this Cubitt raised money from a wide variety of sources. For example, when he did obtain acceptable terms for the south side of Tavistock Square in 1821 he built nine houses for a stockbroker, Benjamin Oakley, who paid him £18,400 in instalments as the work progressed. Most of his schemes were financed by mortgages with assurance companies, banks and private individuals, using completed premises and his workshops as security. Thus, in these respects, even the most successful master builders were similar to their smaller counterparts, speculating on borrowed money, sometimes to the point of near bankruptcy when market conditions deteriorated.

On a lesser scale, provincial towns spawned their own master builders who were capable of completing large and varied contracts. Many of the symbols of civic pride and advancement, such as new town halls and market halls, were put up by these firms – indeed, it was often their expansion to tackle these large public works which set them on the road to success. They do not figure prominently in general histories of the building industry, but local archives are replete with examples of builders who employed teams of regular workers. Examination of the records of some of the most tenacious has shown that they were well organized and had a professional relationship with clients and architects, very much along the lines of modern building firms. One illustration which can be mentioned in passing is the Anelay business in and around Doncaster which from small beginnings in the 1740s grew into a sizeable concern in the nineteenth century. By 1861 its staff had risen to 41 masons, 11 men and one apprentice, and it had completed a long list of substantial contracts, including Doncaster market hall (Stapleton *et al.*, 1975, 14). Up and down the country firms of this sort of intermediate size were making their mark on Britain's towns, partly by speculation but also by competitive tendering for specific contracts.

Support services

Finance

Setting aside civic and institutional contracts, most building activity was not only speculative but financed on borrowed money. The supply of capital to developers and builders underpinned the whole process of urban development and, in general, involved untold numbers of individuals from various walks of life who were prepared to lend their money against the security of the property to be built. Having said that, though, there were those directly engaged in development who were able to fund building work. Wealthy landowners, manufacturers and companies, as well as successful builders, often financed their own schemes and were sometimes willing to make loans to others. On the leasehold Cotton estate in East London, for example, the Trustees gave loans to builders on mortgage, to be repaid when the houses were sold. Further 'builders were supplied with bricks from the estate itself which were meted out as the need arose . . . The system of mortgage advances tided the builder over the period between the start of his operations and the sale of the

Table 11.4 Loans to builders on the Cotton estate, east London, 1864–1879 (K. A. Halstead, 1982, Figure 93)

Date	Number of leases granted	Money outstanding on loan to builders (£)
1864	114	–
1865	136	14,471
1866	112	13,945
1867	166	17,372
1868	110	8,070
1869	100	9,700
1870	72	4,250
1871	56	4,080
1872	63	3,319
1873	39	1,905
1874	109	9,320
1875	136	6,950
1876	109	985
1877	37	350
1878	11	–
1879	2	–

property. Interest was generally charged at 5 per cent but increased to 8 per cent if the loan period was extended' (Halstead, 1982, 55). Table 11.4 shows the financial commitments of this one estate to builders in the period 1864 to 1879 when most of the building was completed.

Likewise, another source of finance from amongst those directly engaged in urban development was from the most successful builders who were prepared to use some of their accumulated capital to back small enterprises. Besides being builders in their own right, these backers also operated as developers and financiers – which once again reinforces the point that individuals played more than one role. One such example was the builder Thomas Cubitt who, in his later years in particular, acted as developer and financier on sites in London which he leased from their owners and then underleased to builders. On the Neathouse estate of Lord Westminster in southern Pimlico Cubitt acted as developer in providing the roads and sewers but most house building was undertaken by small speculators who were indebted to him for their capital and building materials. Typical of the financial arrangements entered into was the following letter sent to Henry Tyler who had taken ground for six houses in Warwick Square in 1851:

The following is something like the way the money for 3 Houses you are building in Warwick Square may be advanced to you on Loan at 5 per Cent.

	£
We have advanced	150
and when 3 pair floor is on & partitions filled	100
when 4 pair floor is on & partitions filled	50
when roof on	100
when slated, gutters on etc. & R.W. pipes	50
when front composed down & Lobby's built	150
	600

The compo, sand & other materials, except Bricks to be deducted.'

(cited in Hobhouse, 1971, 341)

Thus, as a developer, builder and financier, Cubitt not only consolidated his own fortune 'but gave an opportunity to do so to many smaller men. He enabled many small builders and speculators to undertake development, and gave a start to many men who later achieved considerable importance in the industry. Thus Aldin and Harris, well known in the later Victorian period as developers and builders in South Kensington, seem to have made their first ventures on Cubitt's ground in Clapham and South Belgravia in the 1840s. Members of the Trollope family also took sites on Cubitt's ground in the late 1850s from the executors. For a very small builder or an ambitious building tradesman, Cubitt's developments provided a rare opportunity: he could take a well-prepared site in a well-run area knowing that he would be provided with financial backing in the shape of materials and cash as the building progressed' (Hobhouse, 1971, 343). On a smaller scale, this was repeated across the country.

Beyond landowners, builders and others who were themselves directly involved in urban development, most of the capital for building work was provided by large numbers of relatively small-scale investors from towns and the surrounding countryside – local trades people, shopkeepers, solicitors, widows and landlords. Banks and assurance companies did provide capital for particular projects, but it was the private investor who dominated the market. Individually each person invested a small amount of money, but collectively they were the main support for the building industry, both in funding the construction of property and in the purchase of finished houses. On the latter point, it was the general practice for new houses to be purchased by small investors who then let them to tenants, along the lines which Powell observed in Cardiff: 'Buyers of new houses . . . were local people seeking a safe outlet for small amounts of capital and wishing to keep an eye on their assets rather than invest in other places where, at times, returns were higher. Such investors appear hardly ever to have been in short supply and when building flagged it was likely to be due to problems with short-term credit or finding tenants able to pay economic rents. About nine out of 10 houses were let rather than owner-occupied, and ownership was widely spread' (Powell, 1982, 43).

While the rented sector underpinned the speculative housing market, owner-occupation was favoured by some (but by no means all) of the relatively well-to-do who could afford to build or purchase property outright. It was also sought by many of those in steady employment on relatively small incomes who could afford to make regular weekly or monthly savings, and it was these regular savings which provided another source of financial support for builders when people pooled their resources in building societies or building clubs. These building societies were to emerge as national institutions during the nineteenth century but they sprang from quite modest beginnings. Before the 1800s artisans and craftsmen had clubbed together in friendly societies and self-help groups to alleviate hardship among the participants, and it was this spirit of co-operation and mutual aid which was harnessed in newly-formed building societies to provide houses for their members. As Price has written in his history of the movement 'by the last quarter of the eighteenth century the stage was set for housing development and the establishment of building societies. Friendly societies had shown the possibilities and advantages of co-operation; the early years of the Industrial Revolution had witnessed the beginnings of the shifting of the population from the countryside; the Evangelical Revival, in reminding men that they were spiritual beings, had inspired them with the desire, among other things, for home conditions better and worthier than had been known. Building societies were an obvious necessity' (Price, 1958, 13).

From the 1770s onwards building clubs or societies were set up by small numbers of workmen, usually not more than 20, with the specific aim of building houses for themselves. Each society was a savings club into which the members paid a regular subscription; the savings were then used to acquire building plots and to erect houses. In many societies the members themselves performed the building tasks, using materials purchased from the funds; otherwise, builders were employed. Building proceeded as and when funds allowed, and members drew lots for the completed premises. When all members of a club had been provided with a house, and when all liabilities had been settled, the subscriptions ceased and the society was terminated. To safeguard the savings of their members, these terminating societies operated according to strict rules of conduct which were enforced by elected officers. When a member failed to meet his commitments his share was sold off to a new member.

Although the records of most of the early societies have not survived, there is sufficient extant material to reconstruct their operations. One of the well-documented examples is the Longridge Building Society which was active in the small but growing textile community in Lancashire between 6 March, 1793 and 19 January, 1804. It has been described at length by Price (1958, 32–44). Here, 20 members, who were mostly weavers and yeomen, put their names to 'Articles of Agreement for Building a House, with Necessary and Coalhouses, for each Subscriber (according to a Plan to be fixed on by a Majority of the Subscribers)'. Each member agreed to pay the sum of one guinea as a membership fee, and to subscribe the not inconsiderable sum for men of their means of 10s. 6d. on the first Saturday in every month for an unspecified period, until all members had been housed. Members were subject to a set of rules which were enforced by fines against recalcitrants, and even expulsion. Rule six, for example, dealt with some misdemeanours as follows:

> If any Person, during the Club-hours, shall become Quarrelsome, be guilty of Swearing or Cursing, laying Wagers, promoting Gaming, or refusing to keep Silence when desired by any Officer, or shall injure or depreciate any of the Members in their Persons or Character, such Member shall, for every such Offence (to be determined by a Majority of the Members present) forfeit the Sum of one Shilling, or on Refusal of Payment, be excluded the Society, and lose the Benefit of his Subscription-money. Any Officer embezzling the Society's Money, or not making up proper and Satisfactory Accounts, shall be totally excluded the Society, and lose the Benefit of his Subscription. (quoted in Price. 1958, 34).

The progress of this society was impressive, due in no small part to the energy and astuteness of its committee and trustees. Within eleven years the twenty houses had been completed and occupied, and all liabilities had been met, at an average cost per house of £97 8s 6d. Thus, having achieved its original aims, the society was terminated in January, 1804.

By the first quarter of the nineteenth century these terminating societies had become one of the standard means of financing new houses, particularly in the Midlands, Lancashire and Yorkshire. There was no central organization binding them together as a national movement – each was a short-lived, independent body which made its own small contribution to the local housing stock. However, changes were occurring in their practices. It was becoming much less common for societies to build houses; rather, the money was advanced to members either to build their own or to purchase existing houses. Further, under the Building Societies Act of 1836, some national (albeit imprecise) controls were introduced in that their rules had to be approved by a certifying barrister (later given the title of the Registrar of Friendly Societies). Much more notable, though, was the appearance of permanent societies, the forerunners of modern building societies. The idea

Table 11.5 Growth of building societies, 1853–1873 (E. J. Cleary, 1965, Table 4, 46)

Region	1853 Terminating	Permanent	Total	1873 Terminating	Permanent	Total
North	258	15	273	475	234	709
North Midlands	37	7	44	50	36	86
South Midlands	17	1	18	38	21	59
East	14	3	17	35	21	56
London and South-East	390	39	429	266	168	434
South-West	34	14	48	35	32	67
Wales	9	1	10	52	21	73
Scotland	5	1	6	6	2	43
Ireland	1	2	3	2	5	7
TOTALS	765	83	848	959	540	1,534

of permanent societies, usually attributed to an actuary, James Henry James, in a pamphlet published in 1845 (Cleary, 1965, 47) received widespread support and was put into practice. Under the new system, the building society, in return for interest and perhaps bonus payments, received capital from investing members, which in turn it loaned to borrowers who became mortgagors of the society. With prudent management the society could continue for an indefinite period, taking in new investors and borrowers as it grew – in effect, it could operate on a permanent basis.

The second half of the century saw a gradual growth in the number of permanent societies but not to the total exclusion of the terminating, which showed remarkable tenacity in Lancashire, South Wales and the North East. It can be seen from Cleary's figures on the relative strengths of the two types of society (Table 11.5) that terminating clubs were still in the clear majority in the 1870s, with heavy concentrations in the North: 'In Oldham in 1873 there were still 34 terminating societies and only three permanent; and in Rochdale the respective figures were 28 and two. Liverpool had 124 and Manchester 123 terminating societies, though in these two towns the permanents had made substantial progress' (Cleary, 1965, 46). However, at the end of the century, as old terminating societies were wound up, the permanents took the lead, with a few of them dominating the field in terms of assets held..

During the century both types of society provided funds for house building, either directly, by financing building operations, or indirectly, by offering mortgages on property already built by speculative builders. It is clear from the evidence presented to the Royal Commission on Friendly and Building Societies in the early 1870s that the practice varied between societies.

For some, advances to builders was their main business. 'Yes, it is the general practice in our business to advance to builders' (C. L. Gruneison, Conservative Land Society). 'I should think the bulk of our advances are made to builders' (H. J. Phillips, Temperance Society). There were equally strong views on the other side. 'I saw that builders tried to introduce business into it [his society] for the purpose of building . . . and I discountenanced anything of that sort' (J. Ryalls, connected with a number of Sheffield societies). Mr T. F. Peacock was against lending to builders: advances should be made only on houses already built. Lending to builders had its dangers; Mr G. Watson, an Edinburgh solicitor, spoke of 'Men of straw who had never been heard of in the trade, suddenly become builders on a large scale, and after a time disappear (Cleary, 1965, 73).

What is difficult to determine is the overall contribution made by these transactions to house building. On the assumption that the annual amount lent by societies in 1870 was £2.5 million and that the average price per house was around £240, it has been estimated that they were financing the construction of one house in about every seven or eight built (Cleary, 1965, 289). However, while giving a general indication of the relative importance of building societies in house construction, this may be a rather rough-and-ready calculation since the evidence from particular local studies suggests that the overall financial contribution of building societies, either directly or indirectly through intermediaries such as solicitors, was 'considerable if only because of the large number of societies and the size of their business' (Yeadell, 1986, 70).

The middle of the century saw the widespread appearance of another type of society in the form of freehold land societies (though most of them eventually functioned as developers, or as building societies on similar lines to those described above). These societies were first formed with the object of assisting people to acquire voting rights through ownership of property: blocks of land would be bought and then subdivided among the members so that each qualified for a vote (Homan, 1983). However, from the 1840s, while continuing to confer voting rights and attracting support from the main political parties, they were harnessed to buy building land which was then sold off as plots to persons in receipt of loans from ordinary building societies. One of the earliest, that founded by James Taylor in Birmingham in 1847, illustrates their role as developers: 'Taylor continued the operation of the Birmingham Freehold Land Society [after registering it as a friendly society] with his group of normal building societies, the first advancing money to buy the plot of land, the second to erect a house upon it. The land society laid out estates in plots, put in roads and drains and exercised some general control over building; building lines were set, no public houses allowed and the number of shops controlled. At the start plots were allocated to members in rotation, but the considerable success of the society led to long waiting lists and this hampered further progress. To offer a greater incentive to new members, in 1851 a proportion of plots were allocated by ballot and eventually all plots were allocated in this way' (Cleary, 1965, 51).

One further example will serve to show that the same or similar schemes were implemented elsewhere. In Cardiff, a city dominated by 99-year leasehold tenure, the National Freehold Land Society was the developer of one of the few large parcels of freehold land. In 1852 it purchased 110 acres from the Romilly estate in the Canton district for £14,100. Backed by local Liberals, who hoped that their party would receive the votes of the newly-created 40s. Freeholders, the Society issued £30 shares to its members which gave them the right to choose plots. 'The estate was divided into lots 10 feet wide. A shareholder could take any two or more of these lots, depending on the value of the shares held. There was thus considerable flexibility in the size of the plots, from 20 feet upwards, contrasting with the uniformity to be found on leasehold estates. A number of regulations were made: the building line was fixed at 15 feet from the road, and no house was to be worth less than £150' (Daunton, 1977, 81). By 1853 over 100 of these societies had been formed in London and the provinces, most allied to the Tories or Liberals. Although by the mid 1860s they had, to all intents and purposes, lost their political function, they continued to act as intermediaries in purchasing small freehold building plots for prospective owner-occupiers.

Legal and technical services

Finally in this consideration of the decision-makers involved in urban development, brief mention must be made of those who provided legal and technical services. As noted earlier, solicitors were a source of finance, but of course they were also directly engaged in the legal transfer of land to developers and builders and helped to formulate the clauses and covenants which were imposed on building plots. Although it is often difficult to pin down their precise contributions in particular developments, they were always there in the background as sources of advice and guidance. At the broad level it may have been to try to preserve the select character of a neighbourhood for the landowner; at the minute scale it may have been to protect the right of the landowner, developer or builder to construct another house against the gable-end wall of a completed dwelling in a partially built terrace. What is clear from extant documents is that solicitors played a variety of roles in urban development, right from the initial transfer of land, through the raising of capital, to the selling or leasing of completed houses.

During the course of the nineteenth century, the building industry saw the progressive specialization of the main technical occupations associated with land development and house construction. As Powell observed in his discussion on designers and advisers, 'the open and changeable roles and relationships which existed among the providers of new buildings in the first half of the century became less fluid in the second half' (Powell, 1982, 69). The setting up of the Institute of British Architects in 1834 and the Institute of Surveyors in 1868 served to put the activities and dealings of their members on a professional footing. In the case of architects 'the easy-going benign atmosphere of the eighteenth century had gone. Pretensions to artistic ability were no longer sufficient to sell the architect's services; he had also to establish in the public eye his professional reputation. Recommendation was no longer personal; it depended upon public estimation of the profession as a whole' (Kaye, 1960, 83). This meant that architects looked to codes of training and practice to establish their credentials, and divested themselves of other functions, such as quantity surveying and building. Likewise, surveyors set out their own terms of reference with appropriate professional codes (Thompson, 1968). These professionals not only had their separate roles to play in the private market in advising and negotiating for clients, but also in the field of local government where there was a demand for borough surveyors and architects in the new administrations. In local government both professions were seen to have an expertise to offer in the framing of by-laws. For instance, consultation with the professional institutes was written into the Metropolis Management and Building Acts Amendment Act of 1878 which empowered the Metropolitan Board of Works to make by-laws requiring the deposit of plans and regulating the inspection of buildings (Thompson, 1968, 161). By the end of the century architects and surveyors had established their status in the building industry, particularly in the non-speculative and contract sectors of the market and in local government. While perhaps the majority of those engaged in these professions bequeathed a modest legacy to the built form, the most notable architects of the age left an unmistakable imprint through a profusion of architectural styles – Greek revival, Romanesque, Tudor and Elizabethan, Italianate, Gothic – and in a variety of building materials, including cast and wrought iron, glass and terracotta. These styles and materials were applied with energy and enthusiasm in some magnificent public and commercial premises which adorned the districts discussed in Chapters 6, 7 and 8. Since this chapter is primarily concerned with housing, elaboration will not be given here, but it is worth citing just one example, the profound impact of Pugin's advocacy of the

Gothic: 'The variety and ingenuity of neo-Gothicism produced by Pugin's follow-ers is readily apparent in a host of great creations: London's Law Courts, Public Record Office, Prudential Assurance Building and Midland Grand Hotel at St Pancras Station; town halls at Manchester, Congleton, Rochdale, Reading and Northampton and the Theatre Royal at York are random examples. It is no less apparent in innumerable smaller structures such as warehouses, office blocks, market buildings (Columbia Market, Bethnal Green), Institutes (York), working men's clubs, colleges, schools, hospitals, almshouses, fire stations and all manner of buildings, including houses, all over the country' (Burke, 1976, 68–71).

The built form

The end product of the activities of all of those engaged in urban development was a variegated townscape in terms of the quality, style, materials and layout of buildings. Part of that variety was derived directly from the conscious decisions of landowners, developers, builders and others to cater for particular sectors of the housing market; part of it stemmed from the character of the local industrial base, the availability of local building materials and the vernacular building practices in different areas. Given the diversity across the country, and the changes which occurred during the century, many writers on the built form have chosen to examine only particular aspects of the total picture. For example, Chapman (1971) has brought together studies on working-class housing, Muthesius (1982) has concentrated on the terraced house in all its regional forms, and Edwards (1981) has dealt with the design of suburbia. Others have worked on particular features in selected towns or regions, for instance, Forster (1972) on court housing in Hull, Lowe (1977) on industrial workers' houses in Wales, and Smith (1974) on tenements in Scotland. Clearly, each type of housing is worthy of detailed investiga-tion in its own right.

While this variety makes it almost impossible to draw up a comprehensive typology of the urban housing stock in Britain, Burnett (1980) has attempted to summarize the main housing types which emerged in the nineteenth century. He made the broad distinction between working-class and middle-class housing and examined the changing characteristics of their sub-types in England. However, as indicated in the preamble to his description of working-class housing, he was well aware of the difficulty of compiling a classification which picked up all the nuances of type, quality and style within and between towns in Britain: 'Given the contrasts in standards of living between, say, a handloom weaver, a miner and an engineer, and the variations in regional building forms between a London tenement, a Nottingham back-to-back and a Durham miner's terrace, the difficulty is to know what the "norm" was, or, indeed, whether the concept of a norm is useful where abnormality was so typical' (Burnett, 1980, 54). Nonetheless, his is a useful overview and will be briefly summarized here to illustrate the diversity of housing which emerged from the decision-making process set out in this and the previous chapter.

Burnett's scheme was essentially a hierarchy based on quality, embracing the houses of paupers at the one extreme and the villas of the very rich at the other. Towards the lower end of his scale, working-class accommodation in the first half of the century comprised a mixture of premises, of those, on the one hand, which had been adapted from their original use and sub-divided between tenants, and, on the other, of purpose-built workers' houses. In terms of quality, this accommoda-tion ranged from cellar dwellings at the lowest level, through lodging houses, tenements and back-to-backs, to 'through' terraced houses. Superimposed on these broad types were variations in style and layout brought about by local building practices and materials, and by the needs of local industry, particularly in the form of workshop houses. These types persisted in the second half of the century, but the worst of them were increasingly subjected to controls (if not clearance) and new premises were gradually regulated by building by-laws. 'Although by the end of the century the general development of working-class housing was towards more

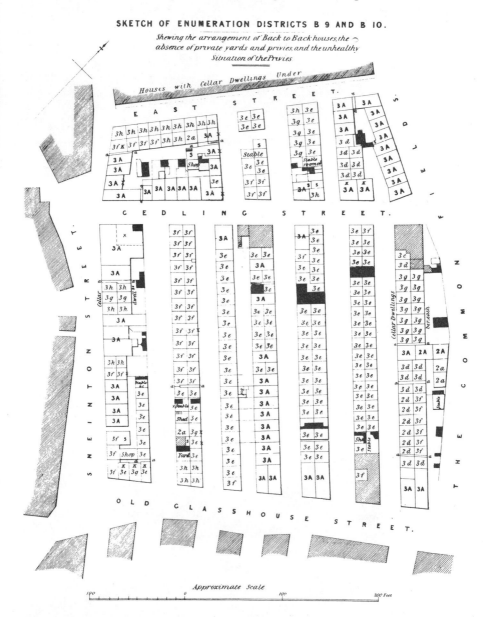

Figure 11.2 Plan of back-to-back houses in Nottingham, 1844.
(after First report of the commissioners on the state of large towns and populous districts, 1844)

Plate 11.1 Tenements at Leith, Edinburgh. Tenement blocks were particularly characteristic of towns in Scotland, but were also adopted elsewhere.

Plate 11.2 Conventional single-fronted small terraced houses in Caroline Street, Newport, Gwent.

Plate 11.3 Woollen weavers' houses in Commercial Street, Newtown, Powys. The two living floors are surmounted by two floors of workrooms which retain their large windows.

Plate 11.4 Silk throwers' houses in Congleton, Cheshire. The two living floors are surmounted by a workroom.

standardized forms, determined largely by legislation, distinct regional types continued to characterize different geographical areas – the through terrace of the South, the back-to-back of parts of the Midlands, Yorkshire and Lancashire, the "up-and-down flat" of Tyneside, the divided, tenemented house of London, and so on' (Burnett, 1980, 152). Some of this variety is illustrated in Figure 11.2 and in Plates 11.1 to 11.4

In his treatment of middle-class housing Burnett accepted the fact that the middle class was not a single homogeneous body but rather a continuum of sub-groups. At its lower end it comprised petty tradesmen, shopkeepers and clerks and merged with the skilled workers of the upper working class, at the intermediate level it included managers, lesser factory owners and members of the professions, and only at the top the large industrialists and merchants. This variety was reflected in middle-class housing. For much of the century, there was little to distinguish the homes of the lower middle class from those of the upper working class. Like their inferiors, they occupied respectable 'through' terraces. They were unable to afford the rents – and the servants – of the fashionable terraces of town houses built for their superiors. When the most affluent sought relief from the disease, noise, congestion and commercial pressures of the city centres, they turned their backs on the town house and looked towards the semi-detached and detached suburban villa. This became the goal of the rising middle-class family, and the building industry responded to this new fashion with the middle-class suburban home. The pace was set by London, but 'by the century's end, every sizeable town exhibited the same phenomenon, though on a smaller scale and at shorter distances from the centre' (Burnett, 1980, 189).

Conclusion

This chapter has taken the process of land development beyond the landowner into the activities of the developer and builder, those who created the built form. However, while developers and builders stand out as key figures in the laying out of streets and the construction of new houses, they were dependent on many others for finance and services. In the context of Springett's decision-making framework, the primary decisions were taken by the landowners, developers and builders, but progress often depended on the support of other interested parties who contributed their money or professional skills.

For purposes of discussion, in this and the previous chapter, the principal figures involved in urban development have been separated out as free-standing individuals, each becoming engaged in development at the appropriate stage. However, as has been stressed from time to time, this is a simplification. While in many instances individuals did perform their own separate functions, it was often the case that they carried out more than one activity and roles became blurred. This merging of roles was particularly pronounced in the first half of the century, as observed by Powell: 'The roles of those who decided to build, whether landowner, developer, builder or investor, overlapped and merged so that within the tangle of aims and methods were great diversity and few universally recognized procedures' (Powell, 1982, 3). At the same time, while containing a growing number of master builders, the building industry was fragmented into innumerable small and often short-lived concerns.

Throughout the discussion reference has been made to speculation and financial gain; those taking part in urban building have been portrayed as the stalwarts of *laissez-faire*, making a profit where and when the market allowed. However, it was this very freedom in the housing market which eventually led to the introduction of controls on urban development. In the absence of building regulations, those in search of a large profit or quick return had cut corners in the construction of premises and in the provision of basic utilities. The overcrowded and disease-ridden slums created by this process of unchecked development spurred those with a social conscience to press for statutory controls. As the century wore on these voices were heard by local and central government, and gradually measures were introduced which were to have a direct bearing on the layout of streets and buildings, on the construction of property, and on the supply of utilities. This will be examined in the next chapter.

12

Controls and constraints

To this point in the book it has been argued that intra-urban geographical patterns were shaped by the forces of demand and supply in the market place. The city has been seen as the product of uncontrolled *laissez-faire*. While this argument contains a large measure of truth, it is necessary to introduce one important qualification to it, to allow for the fact that from the eighteenth century onwards attempts were made by various statutory bodies to impose controls on new building and to regulate or improve the existing urban stock. It is the aim of this chapter to look at the way in which such controls were applied to towns. In part this means looking at the evolving system of urban administration as new bodies were set up to deal with poor living conditions in general, and public health in particular. However, it is not the intention to review all of the intricacies of the administrative system itself, but rather to focus on those bodies and measures which had a direct bearing on the built form on the ground. Although there was no formal town-planning machinery as such, it can be suggested that the innumerable pieces of urban legislation which appeared on the statute books over the course of the century were the foundation stones of the modern town-planning movement in Britain.

So great was the pace and scale of urban growth through industrialization that some control over the internal layout of towns and the provision of basic facilities became inevitable. It is paradoxical that the ideology of the free market, or *laissez-faire*, which fuelled much of this new urban growth and gave hope of advancement to thousands of migrants, at the same time created abominable living conditions for so many inhabitants. On its own, the free market could not, or would not, cope with the stresses in its expanding towns. It was this failure which heralded the rise of town planning as both national and local government were drawn into the urban arena to try to curb the worst excesses of congestion and concentration and to alleviate extreme deprivation. Sutcliffe makes this point effectively in his introduction to the edited volume of papers from the First International Conference on the History of Urban and Regional Planning, '. . . the market proved incapable of regulating the massive conflicts generated within the industrial town by economic forces of unprecedented power and related social divisions within the urban populations . . . Discovering that the problems of the growing towns threatened it more than in the past, the State was inclined to involve itself increasingly in urban affairs, though it was prepared to leave much of the initiative to local administrations generated within urban communities' (Sutcliffe, 1980, 2).

This is not to say that no efforts were made to improve living conditions in

Britain's towns by those who amassed their wealth in the free market. Throughout the century, as well illustrated by the examples described by Tarn (1973), model communities were built by individual industrialists for their workers. Starting with the activities of Samuel Oldknow and Robert Owen at the end of the eighteenth century, there was a succession of entrepreneurs, including Titus Salt at Saltaire, Edward Akroyd at Akroyden, William Hesketh Lever at Port Sunlight and George Cadbury at Bournville, who commissioned what were generally small villages. To these can be added the model housing schemes which were instigated by a variety of philanthropic organizations which appeared from the 1840s onwards, mainly in London. Some were founded on the wealth of principal benefactors, such as the Peabody Trust, supported by the banker George Peabody, and the Improved Industrial Dwellings Company, financed by Sydney Waterlow; others were promoted by zealous reformers, for example the Society for Improving the Condition of the Labouring Classes which included Lord Ashley and Dr Southwood Smith among its supporters. However, too much should not be read into these efforts. Not only were model dwellings generally uninspired in design and layout, being 'either a repetition of part of the existing housing patterns or else rather gaunt tenement blocks' (Tarn, 1980, 81), but their impact on the sharp practices of the ubiquitous speculative builder was slight. While these schemes did set an example, the evidence for the nineteenth century indicates that the speculator responded more to coercion by legislation than to persuasion by example.

Turning to look at control by legislation it is convenient to distinguish between the involvement of local and national government in urban affairs. It is recognized that this distinction is somewhat arbitrary since local government itself was gradually structured by and dependent on the powers conferred by national legislation. However, in the first half of the century at least, a great deal of the initiative against poor urban living conditions was shown by local bodies using powers derived from their own Local or Private Acts of Parliament. The State was drawn in later and was able to benefit from the experience gained from the formulation and implementation of these local measures. In the following discussion it is proposed to concentrate on the pattern of control which evolved in England, but, in general, the same principles applied elsewhere in Britain, even though separate legislation may have been involved.

Local government

As is plainly evident in the attempts by Weber (1899/1967) and Law (1967) to define towns and to assess the size of the urban population in the nineteenth century, there was no administrative tier which encompassed all urban places. On the contrary, a profusion of overlapping bodies – including parishes, boroughs, improvement commissions, boards of health, and sanitary authorities – was responsible for towns. Since each body performed its own particular range of functions for its own particular territory, it is necessary to probe the actions of this miscellany of statutory authorities in order to identify the controls imposed on town development. It was not until late in the century that the administrative system was tidied up when, through the Local Government Acts of 1888 and 1894, urban affairs were placed firmly in the hands of Borough and Urban Councils.

Earlier in the century there had been moves to reform some areas of municipal administration. Most notable was the Municipal Corporations Act of 1835 which was applied to 178 of the 246 boroughs (excluding London). However, on its own, the 1835 Act did not transform the boroughs into effective administrative bodies:

'There was less corruption, but otherwise borough government after reform consisted of a more or less different body of people doing the same things as before in the same way' (Ashworth, 1954, 70). While the Act itself may not have brought about major administrative changes, its timing coincided with the growing interest of the State in urban conditions, and this has led many writers on the history of local government in the nineteenth century to see the 1830s as a watershed in the form and style of administration. 'Up to this time the central government had left the local communities very much to themselves, allowing them to regulate their affairs by local custom or by private and local acts, rather than in accordance with any general national pattern, or concept of how things should be done. But the great age of local act legislation was passing, and public general acts, covering the whole country, were to supersede, to a great extent, the former system' (Keith-Lucas, 1980, 149). Up to this time towns were administered by a number of bodies, primarily the unreformed boroughs or municipal corporations, parish vestries, and trustees and commissioners, each of which performed some specific functions but none of which had the powers or resources to deal with the variety of problems associated with rapid urban growth.

At first sight, out of these bodies the boroughs might appear to have been best equipped to cope with urban problems through their mayors and corporations. However, the boroughs comprised an extremely mixed and rather limited collection of settlements, ranging from and including only some of the great cities – for example, Leeds, Nottingham and Bristol – to 'obscure villages, where no-one could quite remember whether there was a corporation or not' (Keith-Lucas, 1980, 15). Further, while members of the corporation performed a variety of public duties, fortified by wining and dining at the expense of the corporate revenue, in general the corporations themselves did not supervise urban improvement. As Keith-Lucas has shown, direct responsibility passed to other bodies, although some members of corporations served on them.

> Very few of the corporations concerned themselves directly with the business of drains and sewers, paving and lighting the streets, water supply and public health. This does not mean that such matters were ignored. In the early nineteenth century there was a ferment of development and improvement. Great numbers of Acts of Parliament were passed to provide services of this kind, nearly always at the instance of the corporation. But, with only a few exceptions, the powers were granted, not to the corporations themselves, but to specially created statutory bodies – the improvement or paving commissioners. In these bodies the mayor, aldermen, capital burgesses, etc., would commonly have seats on the governing body, along with other leading citizens, *ex officio* or elected. But there were some towns, including Richmond in Yorkshire, Nottingham and Macclesfield, where the corporation itself, under a local paving act, undertook such responsibilities, and there were others, including Leeds, where the vestry undertook these functions as part of its responsibility as the Highway Authority (Keith-Lucas, 1980, 33).

Thus, in only a small minority of cases did the corporations themselves undertake such responsibilities.

Setting aside the parish vestries, which were primarily responsible for ministering to the needs of the poor, urban problems were largely dealt with by the separate boards of trustees or commissioners created by Local Acts of Parliament. Before, and for a long while after the Municipal Corporations Act and the passing of national legislation, these bodies were the main agents trying to provide services and improve the environment in Britain's towns. In their classic study of 1922, the Webbs grouped them all together under the title of 'Statutory Authorities for Special Purposes' (S. and B. Webb, 1922), but in reality they showed great diversity in composition and function. Among these *ad hoc* authorities, those which were referred to as Lighting, Watching, Paving, Cleansing, Street or Improvement Commissions were those most actively concerned with urban living conditions. In the Webbs' assessment, they were the foundation stones of later urban administration: 'The establishment, between 1748 and 1835, in nearly every urban centre, under one designation or another, of a new statutory body . . . was, in fact, the starting-point of the great modern development of town government. And it is these Improvement Commissioners, not the Mayor, Aldermen and Councillors of the old corporations, who were the progenitors of nearly all the activities of our present municipalities' (S. and B. Webb, 1922, 235–6).

These authorities were brought into being in their hundreds to remedy a very large number of deficiencies in both municipalities and unincorporated towns. Initially, the majority had fairly limited aims, but with the passage of time they took an interest in a wide range of urban problems. In general they were empowered to levy a local rate to finance their activities (something which frequently brought opposition from residents). Typical of reasonably successful Improvement Commissions were those set up in Birmingham and Manchester. In Birmingham the first attempt to get a Local Act in 1765 failed due to various objections including opposition to the proposed new rate, but four years later an Improvement Commission was incorporated. Its functions were extended by a subsequent Act of 1773 and during the last quarter of the century it tried to remove obstructions in narrow streets, such as bow windows, stone steps at front doors and cellar entrances, to pave the footways in a few of the principal streets, and to provide some street lighting. But the overall impact on the town was slight. However, at the start of the nineteenth century the Commissioners gained a new lease of life under two new Local Acts in 1801 and 1812 which gave them a higher rate income and the power to borrow. From then on they became a much more effective and efficient municipal government, and, strengthened by a further Act in 1828, they were able to bring about notable improvements in some parts of the city. This success ensured that 'the Birmingham Improvement Commissioners continued in full activity, as a fairly efficient governing body, until they were, in 1851, by mutual consent, merged in the Municipal Corporation' (S. and B. Webb, 1922, 256). In Manchester the first improvement commissioners were formed as Police Commissioners by a Local Act in 1765. Over the next three-quarters of a century, until their absorption in the Town Council in 1842, they became involved with a wide range of matters as their powers and resources were extended by successive Local Acts – with paving, lighting, removing obstructions, scavenging, fire-fighting, water supply, gas, and so on. What is also interesting about the Manchester Commission is that it is one example which well illustrates the differing degrees of representation which occurred on these statutory bodies in Britain: 'beginning in 1765 as a limited number of named persons co-opting their successors; then transformed in 1792 into a body consisting of the whole of a class (i.e. of substantial householders); and finally, in 1828, being reconstituted as a body of elected representatives' (S. and B. Webb, 1922, 258). The hundreds of other commissions which were set up elsewhere performed similar functions to those in Birmingham and Manchester, although they varied greatly in detail from town to town.

In the late eighteenth century and for a good part of the nineteenth century these improvement commissions were the most important statutory bodies concerned with urban living conditions; as noted earlier, the existing town corporations were little involved in such matters. However, the 1830s decade witnessed the beginnings of a gradual change. It was one of the intentions of the Bill of 1835 dealing with the municipal corporations that the reformed town councils should take over the power of the various statutory bodies which had been granted under Local Acts. In the event, the Municipal Corporations Act left the commissions untouched, but in practice mergers with the new town councils took place, sooner or later. Further, such mergers were encouraged by both the Privy Council, when it issued a charter of incorporation to a new borough, and the Board of Health, when it made an order under the Public Health Act of 1848. In due course the town councils secured and administered their own Local Acts, and increasingly they were aided by general national legislation, such as the Towns Improvement Clauses Act of 1847 which brought together the usual standard clauses included in Local Acts and thereby facilitated an easier and cheaper passage for Local Acts through Parliament. Yet the new town councils did not completely supplant improvement commissioners. The latter continued their work in the unincorporated towns and in some municipal corporations, and when the reformed town councils themselves obtained their own Local Acts they sometimes used them to set up new improvement commissioners or trustees to deal with specific urban problems. Thus, after the 1830s, both improvement commissions and the reformed town councils were involved in urban

Table 12.1 Examples of powers obtained through Local Acts of Parliament to improve urban living conditions (after Keith-Lucas, 1953–54, 295)

The prohibition of the building of new houses without drains, which had to be approved by the Local Authority (e.g. St Helens, 1845; Chester, 1845; Burnley, 1846; Wallasey, 1845).

Power for the Local Authority to order the owners of the existing houses to provide drains connecting with the main sewer (e.g. Southport, 1846; Manchester, 1845; Northampton, 1843; Glasgow, 1843; Exeter, 1832).

The prohibition of building houses without privies (e.g. Newcastle upon Tyne, 1846; Burnley, 1846).

Provision of public lavatories (e.g. Chester, 1845; Gorbals, 1843; Southport, 1846; Northampton, 1843).

Prohibition of the letting as dwellings of cellars in courts, or of any cellar of less than a specified height and window area (e.g. Southport, 1846; Wallasey, 1845; Liverpool, 1842).

Provision of public baths and washhouses (e.g. Liverpool, 1842; Glasgow, 1843).

The appointment of Sanitary Inspectors (e.g. Southport, 1846; Burnley, 1846; Manchester, 1845).

Prohibition of building houses in close courts (e.g. Manchester, 1845).

Prohibition of building houses except with at least one room of at least 108 superficial feet (e.g. Wallasey, 1845; Liverpool, 1842).

Minimum height of rooms in new houses 8 feet (e.g. Wallasey, 1845; Belfast, 1845; Liverpool, 1842).

Power to order the cleansing and whitewashing of the houses of the poor (e.g. Chester, 1845; Newcastle upon Tyne, 1846).

Inspection and licensing of lodging houses (e.g. Gorbals, 1843; Manchester, 1845; Southport, 1846).

Provision of public gardens and recreation grounds (e.g. Leicester, 1846; Chester, 1845).

Provision of public drying-ground for clothes (e.g. Chester, 1845).

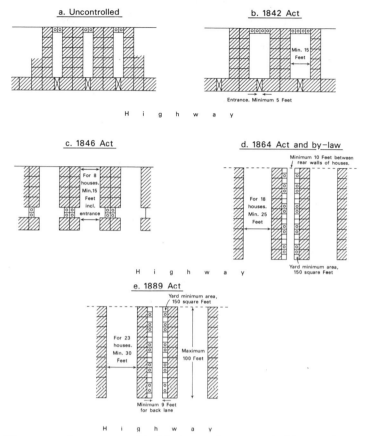

Figure 12.1 Diagrammatic summary of legislation against courts in Liverpool. (based on information in Errazurez, 1946)

improvement. Together they were responsible for very large numbers of Local Acts in the first half of the century. It has been estimated by Keith-Lucas that between 1800 and 1845 in England alone nearly 400 Improvement Acts of one sort and another were obtained. Just before the Public Health Act of 1848 they were particularly active with sanitary improvements, as can be seen from the examples in Table 12.1 (Keith-Lucas, 1953–54, 295).

Given the number and variety of statutory authorities set up by Local Acts it would be a lengthy task to outline representative or typical examples to illustrate the impact of all of them on the ground. Rather than embark on such an exercise it is proposed to do something much less ambitious, to take just two well-documented cases, one in England and one in Scotland, to demonstrate that the local initiatives did work their way through into the built form. Each case deals with a different facet of town improvement, the one with the building of new property and the other with slum clearance and reconstruction. The first example is that of Liverpool which shows the way in which a succession of controls was introduced over a long period to try to eradicate one particular evil in new residential property, congested court housing. The second is Glasgow, which undertook a major clearance and redevelopment scheme at a relatively early date.

Liverpool was one of the first cities to show concern for matters of public health and to translate it into legislation; it sought powers to deal with existing evils and to secure improvements in new buildings. On the latter, Errazurez has set out the campaign it waged against court housing, and it is his study which will be used here (Errazurez, 1946). The city made its first move to improve the environment in 1785 with the Liverpool Improvement Act, followed by another Act in 1825, by which it received powers to carry out street widening in the city centre. These Acts did not control building in new districts and congested court houses (often back-to-back) continued to be built in their thousands on the narrow fingers of land behind the main thoroughfares (Figure 12.1A). It has been estimated that by 1841 nearly one-fifth of the total population of the city lived in courts. However, in the 1840s the city introduced two pieces of new legislation directed at the court, an 'Act for Altering and Amending the Provisions of Certain Acts relating to the Regulation of Building in the Borough of Liverpool' of 1840, and its replacement, an 'Act for the Promotion of the Health of the Inhabitants of the Borough of Liverpool and the Regulation of Building in the Said Borough' of 1842. Under the 1842 Act, generally known as the Liverpool Building Act, the minimum width of courts was fixed at 15 feet, but it allowed entrances as narrow as 5 feet, provided that the height did not exceed 10 feet (Figure 12.1B). The Act stipulated that each house should have access to 'sanitary conveniences', but it allowed one convenience to several houses in courts. Another clause, concerned with living space, fixed the minimum area of rooms on the ground floor as 108 square feet, free of staircases and other obstructions, with a minimum height of 8 feet for any habitable room. However, 'there was nothing in its provisions to compel a modification of architectural types from those of the almost enclosed court, or of the long row of back-to-back houses, with access from a pedestrian passage 15 feet wide' (Errazurez, 1946, 60).

Shortly afterwards, the 1842 Act was replaced by the Sanitary Act of 1846. While confirming many of the provisions of its predecessor, it introduced two important modifications relating to courts. Although it allowed the construction of courts having a minimum width of 15 feet, it prohibited the building of more than eight houses in each such court and required an increase of one foot in width for each house included above that number. Also it required that the entrances to the courts should be the same width as the courts themselves, throughout their whole height. For almost two decades following this Act the usual practice of builders of this type of accommodation was to construct courts of eight houses, with the privies grouped at one or two places (Figure 12.1C). The next move against the enclosed court was the Liverpool Amendment Act of 1864. Section 33 of this Act required that any new court should be open to a public highway at each end, unless it had a minimum width of 25 feet throughout its whole length, including the entrance. Further, it retained the provisions of the 1846 Act on the increase in the width of courts in proportion to the number of houses, which meant that an enclosed court with a width of 25 feet could not contain more than 18 houses and would have to be increased by 1 foot in width for each additional house. Clearly, under these regulations there was little to be gained from enclosed courts by the speculative builder since the minimum width was near that of ordinary streets. The opening up of the courts at either end also encouraged the provision of a privy in each house since the practice of grouping privies at the 'blind' end was no longer possible. However, by the 1860s it was already common to add a little back yard to each house to ventilate the kitchen and house a privy, and this practice was regulated in a by-law of 1864 which fixed a minimum area of 150 square feet for each yard, and a minimum distance of 10 feet between the rear walls of houses (Fig. 12.1D). The final action against enclosed courts was included in the Liverpool Act of 1889. This stipulated that no new court should be built which was not open at both ends to a public highway unless it was precisely 30 feet in width and did not exceed 100 feet in length. It also confirmed that any building would be subject to the Act of 1864 and the relevant by-laws. 'The fixing of a width of 30 feet for courts meant that, in future, no more than 23 houses could be built in each court, which could be done only with the greatest difficulty in a length of 100 feet, with 11 or 12 houses on either side. In fact, the old court, as an internal common space, was abolished and became a short communicating street between two traffic arteries' (Errazurez, 1946, 66). What this Act did was to promulgate the basic type of working-class housing which was built in huge numbers in Britain in the late nineteenth century – terraced houses fronting streets of at least 30 feet width, with back projections into enclosed yards of at least 150 square feet in area (perhaps with separate privies) and back lanes of at least 9 feet width (Fig. 12.1E).

In contrast to this Liverpool example, which has illustrated the imposition of controls on the builders of new property, the second example of Glasgow shows the use of legislation to clear and restructure existing slum districts. At the middle of the century Glasgow displayed the characteristics of those pre-industrial towns which were entombed within expanding industrial cities. 'There was in the centre of this rapidly expanding conurbation a medieval city covering about 100 acres with narrow, winding, ill-paved streets which were quite unsuited to the requirements of an industrial city' (Allan, 1965, 603). Faced with severely overcrowded slum housing and an obsolete street layout, the City Fathers embarked on comprehensive clearance under the City of Glasgow Improvements Act of 1866, which was the largest single undertaking of this kind in the nineteenth century. 'This Act set up the City of Glasgow Improvements Trust with the Lord Provost, Magistrates and the Council as the Trustees. The Trust was given powers to buy up (by compulsion where necessary) and knock down 88 acres of Central Glasgow including practically the whole of what was Glasgow in Adam Smith's time, to form 39 new streets and alter 12 old ones, to sell, lease or feu the cleared lands and to buy ground for a public park at a cost of not more than £40,000. To finance this scheme, the trustees were empowered to levy a rate of 6d. for five years, and 3d. for 10 years and to borrow up to £1,250,000' (Allan, 1965, 604). Initially the Trust was involved in the purchase of property, the displacement of residents and the clearance of sites by demolition, but from the late 1880s onwards, when private builders failed to take up all of the vacant land, it proceeded to build artisans' dwellings, basically employing the tenement block of homes with shops on the ground floor. Over the closing decade of the century it completed the task of demolition which it had started under the 1866 Act and supervised the municipal provision of substantial workers' houses on cleared sites.

Except perhaps for the scale of the scheme, there was nothing particularly exceptional about the Glasgow example. Local powers were widely used to clear slums and to construct new thoroughfares. In nearby Edinburgh, for instance, in the city itself and in the adjacent port of Leith, the town councils secured legislation which allowed them to demolish slum property and re-align streets. The Edinburgh Improvement Act of 1867 'specified, in all the detail of metes and bounds description, the properties that the Corporation was authorized to acquire and clear, embracing, in total, about 3,250 dwellings and various other types of property spread over 34 separate clearance areas. The Corporation was also authorized to use some of the cleared land for opening new streets, or widening existing ones, but the remaining land could be disposed of as it saw fit' (Smith, 1980, 104). The clearance areas are shown in Figure 12.2. Leith was slightly different; while the Leith Improvement Scheme Confirmation Act of 1880 was based on local initiative, by this date the council was able to draw on general national legislation to support its scheme (Figure 12.3).

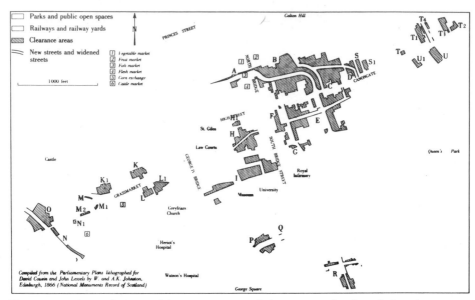

Figure 12.2 Areas designated for acquisition and clearance under the Edinburgh Improvement Act, 1867.
(after Smith, 1980, 107)

It is clear from these selected examples that the practice of securing Local Acts to deal with urban problems continued throughout the century. However, changes did occur in the relative importance of the various administrative bodies responsible for them; in particular, after the Municipal Corporations Act of 1835, such local measures were increasingly sponsored and implemented by the town councils themselves, not the improvement commissions, which were slowly whittled away. In addition to signalling a reduction in the number of improvement commissions, the 1830s heralded yet another and more significant change which was to have far-reaching consequences for towns in Britain, the greater involvement of the State in urban affairs with a view to introducing general legislation which could be applied to all places. From the late 1840s onwards this concern was translated into numerous Acts, many of which were to have a direct bearing on the built form. The next section will summarize the main landmarks in this response by the State.

National government

By the 1830s there were many pointers that the national government would eventually be moved to bring in general legislation to control the most obvious excesses of the free market, and most notably those which had led to the poor state of health and sanitation in the most crowded districts of cities. First, there was some doubt over the achievements of the improvement commissions. While the majority of commissioners and their appointed officers were hard-working and dedicated reformers, a question-mark hung over the overall effectiveness of their efforts in controlling the worst conditions in Britain's towns. In one of their main objectives, that of regulating streets to ensure easy passage, they achieved a measure of success in so far as they did improve the thoroughfares and footways with stone paving in parts of towns. However, they paid little attention to back-streets and slums, and on the wider front of public health and sanitation it has been argued that their record in the early decades of the century was weak:

> even in the best regulated towns, whole streets – sometimes whole districts – remained unpaved, whilst the thousands of densely populated courts and alleys, not to mention the backyards, were usually entirely outside the jurisdiction of any paving authority. These unpaved areas, left in a barbarous state of holes and heaps, became not only the receptacles for stagnant water, but also the dumping grounds of every kind of impurity, which spread, in poisonous dust or liquid filth, throughout the whole district. Thus it came about, in spite of all the work of paving authorities, that when in 1831 the Asiatic cholera reached England, it found actually a larger superficial area of unpaved surface in the midst of crowded human habitations than had existed at any previous period' (S. and B. Webb, 1922, 315).

Second, it was evident that it would be both cheaper and quicker for local authorities to obtain powers if the national government introduced general legislation which could be drawn on by all places as and when required. The fact that towns had to pay their own costs for Local Acts was a disincentive, something which was commented upon by an assistant commissioner of the Royal

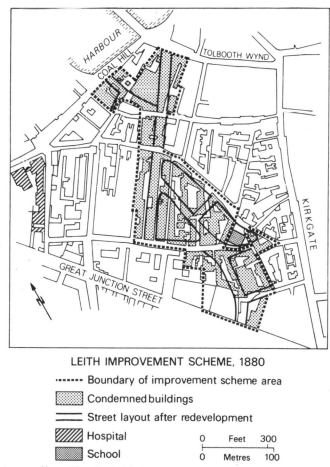

LEITH IMPROVEMENT SCHEME, 1880

∙∙∙∙∙∙ Boundary of improvement scheme area

▨ Condemned buildings

══ Street layout after redevelopment

▨ Hospital

▨ School

0 Feet 300
0 Metres 100

Figure 12.3 Areas affected by the Leith Improvement Scheme, 1880. (after Smith, 1980, 116)

Commission on Municipal Corporations in the 1830s: 'It is surely a hardship that the inhabitants of each town should be compelled to purchase at their own charge one code of laws at least; to buy separately in each particular instance regulations, that are not of private but of public concernment, and that are, or ought to be, applicable to every town in the kingdom, as the matured result of general legislation'. In the event of State intervention, one of the first pieces of general legislation was the Towns Improvement Clauses Act of 1847 which eased the burden for individual towns.

Third, and perhaps most pressing, generally high mortality rates and outbreaks of infectious disease in urban districts spurred the State to take action. In the first instance what action there was consisted of basic fact-finding exercises by various official bodies into the living conditions of the urban poor, and it was this evidence which was put before the central government. For instance, in 1837 Dr William Farr was appointed to the newly created office of Registrar-General and he was charged with the task of compiling vital statistics on the age distribution at death and on variations in the incidence of disease and mortality rates. His annual reports to Parliament highlighted the appalling death rates in towns, and in the poorer districts in particular. 'In Preston, for example, it was pointed out in the 1840s that "the deaths of infants in Preston under one year were – in well-conditioned streets, 15 deaths to 100 births; in middling-conditioned streets, 21 deaths to 100 births; in ill-conditioned streets, 30 deaths to 100 births; and in worst-conditioned streets, 44 deaths to 100 births"' (Briggs, 1968, 230). At the same time as the Registrar-General was embarking on his surveys, the Poor Law Commission was compiling data on health and sanitation. The reports of the Commission in 1838 and 1839 were notable for their studies by Drs Kay, Arnott and Southwood Smith on the fever-stricken districts of London, but most important was the national survey organized by Edwin Chadwick, Secretary to the Poor Law Commission, which was published in 1842 in the classic document *Report on the sanitary condition of the labouring population of Great Britain* (Flinn, 1965). What is most relevant to the discussion here is not the mass of detailed material Chadwick was able to assemble through his army of Poor Law Commissioners, local doctors, ministers, and other knowledgeable persons, but the impact the report had on the government. In Flinn's assessment, Chadwick was the central figure who brought together the various strands in the incipient public health movement and who made the most effective case for general legislation. 'Although its roots stretch back 50 years, the movement was, before 1838, unorganized, leaderless, and, in a legislative sense – the only sense that mattered in the long run – aimless. Essential foundations had been laid, preconditions established, but, important as these were, effective action was missing. This was what Chadwick supplied' (Flinn, 1965, 35).

By the second quarter of the century, then, a variety of pressures were bearing down on the national government. However, general legislation did not immediately follow the publication of the Chadwick report. The first step was another fact-finding exercise, but this time instigated by the State. In 1843 Peel's government appointed a Royal Commission to investigate the health of towns under the chairmanship of the Duke of Buccleuch. Its findings, published in two reports in 1844 and 1845, corroborated the growing mass of evidence on the poor state of urban sanitation and health and supported the case for general legislation advocated by Chadwick and his contemporaries in the newly founded Health of Towns Association. 'The Royal Commission of 1844–45 concluded that no rapid improvement in the condition of the buildings in the most densely crowded districts could be expected but that it would be comparatively easy to prevent the recurrence of similar evils in future. It recommended new legislation to lay down general regulations about buildings and street widths, to bring common lodging houses under public inspection and control, to place the necessary arrangements for drainage, paving, cleansing and ample water supply under a single administrative body, . . . and to give the central government power to inspect the execution of all general sanitary regulations in large towns' (Ashworth, 1954, 53). In effect, the scene was set for legislation. It soon followed in the Public Health Act of 1848 (although it should be noted that in Scotland a Public Health (Scotland) Bill was rejected in 1849 and it was not until 1867 that the Public Health (Scotland) Act finally gained approval).

While the Act was a great achievement in that it did place public health on the statute books, it fell far short of introducing strong mandatory controls. 'The machinery of public health control was set shakily in motion, but Chadwick failed to convince the government that an element of compulsion must be introduced' (Gauldie, 1974, 133). The most important achievement of the Act was the creation of a new General or Central Board of Health which in turn approved Local Boards of Health with their own medical officers, but there were deficiencies in the procedures laid down for the establishment of the Local Boards and the powers attached to them were permissive, not compulsory. The Local Boards, which in incorporated towns were to be the town councils, 'were only to be compulsory in places where the death rate exceeded the arbitrary figure of 23 per 1,000; elsewhere, local boards could only be established upon petition from not less than one-tenth of the ratepayers. Thus, the system of local boards could never hope to be more than partial. Nor were the local boards to be required to appoint medical officers; they were merely permitted to do so if they wished. Again, the local boards were given powers to undertake any necessary cleansing, paving, sewerage and water supply, but were not required to provide these services' (Flinn, 1965, 71). Despite these drawbacks the 1848 legislation was another weapon which could be used in towns, alongside the powers conferred by Local Acts. In the following decades Local Boards were set up in very large numbers. Briggs has estimated that between 1848 and 1854, when Chadwick was a member of the General Board, about 182 localities adopted the Act, and hundreds more followed over the next twenty-five years or so, during which period the General Board itself was dissolved and its supervisory functions transferred elsewhere under the provisions of the Local Government Act of 1858. Spurred on by the reports and recommendations of their own officers, these boards became involved in a great number of specific duties, including sewerage, drainage, water supply, management of streets, maintenance of burial-grounds, and the control of offensive trades. While the 1848 Act did not apply automatically to all places, it is quite clear from the administrative histories of many towns that they achieved some success locally – for instance, through the application of by-laws – in dealing with poor living conditions. Further, and more important in the long run, the Act set the scene for subsequent national legislation and State intervention in urban affairs. In passing it can be noted that what this Act also did, of course, was to add another layer of Local Boards to the already confused and overlapping system of urban administrative bodies which was carried forward into the second half of the century. This proliferation of bodies is something which can be appreciated from the inventory of urban authorities compiled by Dr William Farr, head of the Registrar-General's Statistical Department: 'Dr Farr listed in 1873 198 Parliamentary boroughs (including 13 Parliamentary constituencies, comprising 58 Welsh boroughs) and 224 municipal boroughs. Of local board districts there were 721; these comprised 146 municipal boroughs and 575 towns with a local board but no municipal institutions. Besides, 88 towns had bodies of improvement commissioners of different kinds; 37 of these were also municipal boroughs, 51 having no municipal constitution. Ninety-six towns – as defined by the Registrar-General – were without any municipal body, improvement commissioners or local board of any kind' (Lipman, 1949, 81–2).

In terms of national legislation the middle of the century was something of a turning point. Following the 1848 Public Health Act, and the roughly contemporaneous Nuisance Removal Act of 1846, there was a steady increase in the amount of sanitary and housing legislation, and a growing concern over the chaotic state of urban administration. Table 12.2, which has been compiled from summaries

Table 12.2 Summary of general urban legislation in the second half of the nineteenth century (after Dyos, 1968, 152 and 194)

1846   Nuisances Removal Act. The earliest of a series – 1853, 1863, 1866 – progressively defining conditions unfit for living accommodation.

1848   Public Health Act. Tentative, short-lived provisions, but first acceptance by the government of responsibility for health of the population.

1851   Common Lodging Houses Act, Labouring Classes Lodging House Act. For the regulation of the former and establishment of the latter. Known as Shaftesbury Acts after their promoter. Very limited application.

1855   The Labourers' Dwellings Act. Incorporation of public companies for providing improved dwellings for the working classes.

1866   Sanitary Act. Sanitary Inspectors made compulsory in urban areas; overcrowding classified as a nuisance.

1866–7   Labouring Classes Dwelling Houses Act. Public money first made available through Public Works Loans Commissioners for public companies, housing associations and private persons building dwellings for the working classes.

1868   Artisans' and Labourers' Dwellings Act. Named after McCullagh Torrens, who promoted it as a private member and probably inspired by achievements of improvement acts in Glasgow, Liverpool, Edinburgh. Local authorities in places with over 10,000 inhabitants empowered, but not compelled, to force owners of insanitary dwellings to demolish or repair at own expense.

1872   Public Health Act. Appointment of Medical Officers of Health made compulsory in urban areas.

1874   Working Men's Dwellings Act. Municipal corporations able to make grants or lease land for such dwellings and meet expenses out of rates. Hardly used.

1875   Artisans' and Labourers' Dwellings Improvement Act. 'Cross Act No. 1' permitted local authorities to make for the first time plans for rebuilding slum areas; rehousing to be carried out on the spot. Birmingham leads the way with first clearances for Corporation Street.

1875   Public Health Act. A consolidating act that provided nation-wide administrative machinery for sanitary measures under the supervision of the Local Government Board (created 1871). Local authorities now able to make building by-laws. Cellar dwellings banned.

1879   Second Torrens and Cross Acts, Public Works Loans Act. Mainly concerned with compensation, rehousing and easier public loans. Collectively implied acceptance by parliament of a policy of housing reform though not of municipal management of houses.

1882   Municipal Corporations Act. Eased conditions of repayment of loans to Exchequer, and made it easier for municipal land to be used for working-class housing.

12.2 *Continued*

1882   Artisans' Dwellings Act. Powers to demolish sound buildings that obstructed slum clearances.

1885   Housing of the Working Classes Act. Consolidated the Shaftesbury, Torrens and Cross Acts, eased loan conditions, closed some sanitary loopholes.

1890   Housing of the Working Classes Act. Nothing new: an attempt at simplifying procedures and encouraging both municipal and private enterprise in supplying working-class dwellings; heavier penalties for not complying with closing orders. Led to more widespread use of legislative powers throughout the country.

1894   London Building Act. The first thoroughly effective building code, which was later widely followed.

1899   Small Dwellings Act. First legislation giving local authorities powers to lend money to householders to buy their houses.

1900   Housing of the Working Classes Act. More simplification of housing procedure by municipalities who are now given powers to set up lodging houses outside their own districts. LCC estate at Tooting follows.

published by Dyos (Dyos, 1968, 152 and 194), sets out those measures which were designed to improve urban living conditions in general, and the lot of the slum dweller in particular. In this list the most important piece of legislation for the built form in the second half of the century was the 1875 Public Health Act. There was a strong case for this Act and its associated legislation on at least two counts. First, it was clear by the late 1860s that the voluntary or permissive principle embodied in existing powers was no longer tenable, and that firmer central control was required. Despite the actions taken by local Boards of Health under the terms of the 1848 Act and various local Acts, there is no doubt that in many towns the arrangements for the disposal of sewage and refuse, the supply of fresh water and the control of building standards remained poor. In Leeds, for example, where the council had obtained a local Act in 1842 which was 'one of the most comprehensive and complete which had then been obtained by local authorities', the sanitary condition continued to cause concern, to the extent that 'in 1858, 1865, 1870 and 1874 the town was visited by officers of the Medical Department of the Privy Council, and their findings led John Simon to declare unequivocally that the administration of public health "in proportion to the importance of the town may perhaps be deemed the worst that has ever come to the knowledge of this department"' (Barber, 1980, 302). The second reason for new legislation was that of tidying up the existing profusion of urban bodies by placing local government in the hands of a single administrative authority. These points were brought out in the report of the Royal Sanitary Commission, 1868–71, which recommended new legislation to create a strong central body, to reform urban administration and to revise and codify all the laws relating to public health. By 1875 each of these had been achieved. In 1871 a central Local Government Board was established; this was followed by the creation of Urban Sanitary Authorities; and the great Public Health Act itself contained all the laws relating to public health.

On the administrative front, the legislation of the early 1870s certainly tidied up the arrangements within towns, and the relationship between towns and the State. 'Local government in England had now got for the first time a comprehensive and fairly intelligible system of administrative authorities suited to modern requirements. Town Councils, Local Boards and Improvement Commissioners (the last-named bodies dwindling in number year by year) were henceforward the "Urban Sanitary Authorities" for the administration of the Act of 1875 ... And at the head and centre of the whole new system stood a well-equipped department of government as administrative authority for the supervision and control of all local administration' (Redlich and Hirst, 1970, 159–60).

On the health and housing front, under the 1875 Public Health Act, the Urban Sanitary Authorities were empowered to make and apply building by-laws to new residential developments to ensure that at least certain minimum standards were adhered to by speculators. Although at first comparatively few places used this facility, the position gradually changed, and by the end of the century by-law housing had become one of the most distinctive features of urban expansion. In Ashworth's assessment 'nothing else made so much difference to the physical appearance and condition of British towns' (Ashworth, 1954, 91). The die was cast by the central Local Government Board when it issued a set of model by-laws relating to sanitary and housing conditions, covering such matters as the layout and width of new streets, the construction of new buildings, the provision of space around premises, ventilation and sanitation. It was the progressive adoption of these standards by urban authorities which brought marked improvements but which at the same time stamped what was to become a rather monotonous appearance on the physical form of acres of new working-class suburbs. 'Typically, the "by-law housing" which spread over large areas of working-class suburbs in London and provincial towns in the late nineteenth century consisted of repetitive terraces of four, eight or more houses, intersected by passages or tunnels ("ginnels" in the North) which gave access to small, walled private yards containing a privy and, perhaps, a coal-house, and set in long, parallel, treeless streets from whose pavements the front doors usually opened directly. In those of a better class there might be a tiny front garden with palings to separate the house from the pavement, a bay-window and a small rear garden in place of the paved or cemented yard, but in either case the outstanding characteristic was that of a through terrace house with a "tunnel-back"' (Burnett, 1980, 156–7). The typical tunnel-back terrace house built for skilled and better-paid workers (Figure 12.4) offered reasonably comfortable living conditions. 'Internally, the effects of by-laws also brought important gains. Sounder constructional methods and materials meant better insulation from cold, damp and noise; higher ceilings (usually a minimum of 8ft 6in) and larger window openings gave greater light and ventilation, especially in bedrooms; timber flooring increasingly replaced stone flags; staircases became less steep and tortuous; and there was better provision of fireplaces, sinks, coppers and iron cooking-ranges which usually combined an oven and water-heater. By the 1890s individually piped water supplies to working-class houses were common in many towns, though not yet universal, and gas for lighting and cooking on the penny-in-the-slot system was making much progress. Finally, the through terraced house almost invariably provided a considerably larger floor area than its predecessor' (Burnett, 1980, 158). Rows of these terraces with their single- or two-storey rear extensions were the most significant achievement of State intervention in late century urban expansion; they were the culmination of decades of protest and pressure by sanitary reformers.

Thus far in the discussion, attention has been drawn to the progressive involvement of local and national government in matters relating in particular to health and sanitation in towns. However, government intervention in urban affairs occurred on a much wider front than this, and led to other improvements in living conditions and the environment. Of particular importance was the growing municipal involvement in the supply of public services and utilities – through what Falkus has described as 'municipal trading' (Falkus, 1977) – and in housing.

Public utilities and services

Local authorities extended themselves into a very wide variety of commercial or trading activities for the benefit of their communities: '"Trading" is a loose term; under various definitions it may include not only the four basic local activities of water, gas, electricity and tramways, but also a host of others. Many local authorities ran markets, erected public baths and washhouses, undertook various

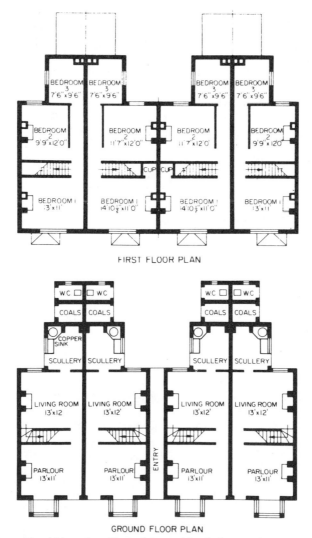

FIRST FLOOR PLAN

GROUND FLOOR PLAN

Figure 12.4 'Tunnel-back' housing. Typical 'tunnel-back' house, the post-1875 solution to mass-produced, high-density, low-cost, working-class housing: narrow frontage, minimal back yard, but through-ventilated and with a water-closet and cold-water tap in each. (after Burke, 1971, Figure 95, 142)

housing schemes, and controlled parks and burial grounds. A few showed greater initiative. By the end of Victoria's reign Sheffield corporation had established a municipal pawn shop, Bradford was running a conditioning house for wool products and a depot for sterilized milk, Bournemouth owned a municipal golf links and Brighton an aquarium' (Falkus, 1977, 135). In part they installed their own facilities, but in many cases they purchased and then enhanced the undertakings of private companies.

It is evident from municipal records that such enterprise did not meet with universal approval. The spirit of *laissez-faire* was very much alive. In Manchester, for example, where the Police Commissioners promoted their own gas works in 1817, some local capitalists saw this as an attack on private enterprise at public expense and as an undesirable monopoly: ' "It was forgotten" said a vigorous local critic, "that, whatever right the Commissioners might have to light the streets with gas . . . they had no more right to monopolize the manufacture of gas for the lighting of private establishments than they have to monopolize the spinning of cotton wicks . . . At great expense we have converted the Commissioners of Police into dealers and chapmen, and have secured to them the exclusive sale of their wares at such price as they shall deem reasonable" ' (quoted in S. and B. Webb, 1922, 264–5). Then again in Leeds, where the council became involved in the supply of all of the utilities in the second half of the century, some opposition was voiced on the grounds that this was municipal interference in what were considered to be the rightful affairs of private entrepreneurs. When the council proposed to purchase the waterworks in the early 1850s 'there were councillors who looked askance at the proposal for municipalization. One alderman, for example, urged that "there was a broad distinction to be drawn between the supply of water to a town and administering the ordinary duties of a town council, such as sewerage, police matters, lighting, paving, etc. A parallel had been attempted between the water supply and the sewerage; but there was this wide distinction that the idea of profit had never been entertained as derivable from the construction of town sewers" ' (Barber, 1980, 317). Despite this opposition, the second half of the century saw a steady increase in municipalization, and this process was aided by general Acts of Parliament.

A good example of the municipalization of the main public utilities is Leeds, where the council acquired the waterworks in 1852, the two gas companies in 1870 and the tramways and the electricity supply in 1894 and 1898 respectively (Barber, 1980, 302). In the case of the waterworks, the town council became involved in its management as early as 1837 when a new joint-stock company was formed, with a board composed of equal numbers of shareholders' directors and councillors. Furthermore, under the 1837 Act the council gained the right to buy out the company at any time after 12 years had elapsed. By the early 1850s the company was sorely in need of additional water supplies to meet the growing demand but lacked capital, and it was at this point that, despite some local opposition, the council purchased the undertaking for £227,417 in November 1852. From then on, but with caution at first over its financial commitment, the council obtained its own Acts to increase its supplies; by Acts of 1856 and 1862 it gained permission to pump water from the river Wharfe, and then it went for a much more ambitious scheme when under the Leeds Waterworks Act of 1867 it was given approval to impound the Washburn in a series of reservoirs along its valley. Over the decade 1869–79 the council constructed the Lindley Wood, Swinsty and Fewston reservoirs at a cost of £508,173, which ensured that 'by the 1880s . . . planning and investment in the waterworks had completely overcome the effects of the myopic policy of the 1850s' (Barber, 1980, 318).

No sooner had the reservoirs been put in hand than the council turned its attention to the supply of gas. Although councillors had discussed the possibility of manufacturing town gas on a number of occasions, up until the end of the 1860s it was supplied by two private companies, the Leeds Gas Light Company dating from 1818 and the Leeds New Gas Company of 1835. In part spurred on by the fact that not one but two companies were digging up the streets to lay pipes, the council eventually acquired the undertakings by an Act in 1870 at a cost of £763,245. In the same year the council gave support to a private application to construct and run a tramway, having initially favoured building and operating its own system, as expressed by Alderman George:

> if there was a great benefit derived from laying down tramways, and if there was to be great profit derived from tolls from the passage of carriages, the inhabitants of Leeds ought to have the benefit. The Corporation had spent £60,000 in the purchase of the tolls, and had expended a good deal of money in the improvements of the streets; they were at present asking Parliament for powers to take over the gas companies, so the streets might be under no control but their own, and it would be hardly consistent to allow the promoters of these tramways to assume the powers over the highways which they wanted. (cited in Barber, 1980, 320)

The private tramway was built, but it was subsequently taken into municipal ownership in 1894 at a cost of £112,226. In the supply of electricity the same pattern was followed. Having considered setting up its own electric lighting company in the 1860s, the council put its weight behind a private venture, the Yorkshire House-to-House Electric Company, in 1891. Before the end of the decade it regretted its initial reticence and expended £217,420 in 1898 to buy out the company.

While private companies continued to be formed in large numbers and were able to maintain a substantial share of the total market, the pattern in Leeds was repeated across the country. By the 1870s the municipalization of utilities was firmly established, as can be seen from the trend in Table 12.3 which shows the adoption of water and gas supplies in the United Kingdom before 1914. The motives of these urban authorities were rarely clear-cut; usually their involvement stemmed from a variety of factors, including dissatisfaction with the state of existing amenities, concern over the prices charged by private companies and the desire to increase their own revenue. What also shows through in many of their schemes, such as reservoirs and public buildings, was civic pride.

Table 12.3 Number of Municipal Corporations adopting water and gas supply before 1914 (after Falkus, 1977, 152)

Date	Water	Gas
Before 1845	10	6
1846–55	29	10
1856–65	22	12
1866–75	66	38
1876–85	68	44
1886–95	42	18
1896–1905	69	37
1906–14	20	11

Housing

Municipal involvement in utilities and services also extended to the field of housing, though the powers conferred on authorities by the State were not widely taken up until very late in the century. As early as the Lodging Houses Acts (Shaftesbury Acts), 1851, it was legislatively possible for councils to build houses for the poor. 'It allowed that income could be provided by boroughs either out of the borough funds, or by raising a rate for the purpose, or by borrowing on the strength of a mortgage of the rates, or by borrowing from the Public Works Loan Commissioners, for the purchase or renting of lands, the erection or purchase and repair of buildings suitable for lodging houses, their fitting and furnishing' (Gauldie, 1974, 241). However, apart from Huddersfield which had housed 2,870 people by November 1864, the Act was ignored in most towns. These opportunities were consolidated by the Labouring Classes' Dwelling Houses Act, 1866, which permitted local authorities to obtain loans from the Public Works Loan Commissioners to purchase sites and erect dwellings. Again the reception by councils was poor, and only one local authority was granted a loan, the Corporation of Liverpool, which received £13,000 for the building of St Martin's Cottages in 1868. This interest by Liverpool in public housing surfaced later when it built two more blocks of artisans' dwellings on its Nash Grove site in 1885 and 1891.

The legislation of the decade which followed the generally unsuccessful 1866 Act, most notably the Artisans' and Labourers' Dwellings Act (Torrens Act), 1868 and the Artisans' and Labourers' Dwellings Improvement Act (Cross Act), 1875, also facilitated slum clearance and the building of labourers' dwellings for later resale, but according to the evidence in the Royal Commission on the Housing of the Working Classes, 1884–85, between the passing of the Cross Act and 1884 only nine towns had begun improvement schemes, and only four towns were replacing the houses they demolished (Gauldie, 1974, 278–9). One of the features of this phase of municipal involvement was the tendency to build new properties with much higher rentals than could be afforded by the displaced tenants. Typical was the scheme approved in Walsall where conditions were such that 'the evils connected with such houses, courts, and alleys, and the sanitary defects in such area cannot be effectively remedied otherwise than by an improvement scheme for the rearrangement and reconstruction of some of the streets and houses within such area' (Royal Commission on the Housing of the Working Classes, 1885, 801). Following the compulsory acquisition and clearance of slum buildings, it was intended to construct a new public road, to widen some of the existing streets and then to release building plots. Having gained approval the local authority implemented its scheme: it demolished 120 houses 'of the most wretched description', set out the new streets and offered building plots, but as was the case in so many improvement schemes, the rents of the new houses were too expensive for the former residents – 'Not any of the tenants disturbed by the improvement scheme occupy the new houses. The rents are too high for them' (Royal Commission on the Housing of the Working Classes, 1885, 801). While action such as this improved some of the most congested areas in a few cities, at the same time it may well have helped in the creation of others by decanting displaced residents into them.

In the wake of the Royal Commission's recommendations, the Housing of the Working Classes Act, 1890, brought together all previous useful housing legislation and paved the way for progressive local authorities to begin development schemes. Yet even at this late stage in the century 'it allowed for, but did not encourage, the actual building of houses by the local authority (as opposed to the letting of ground to builders for the purpose of providing housing). It did not envisage continuing ownership of housing by local authorities, providing that where building was carried out it should be sold or disposed of to private owners within 10 years' (Gauldie, 1974, 293–4). Despite this attitude, the pace of council-house building and ownership quickened, with the London County Council in the vanguard with its Boundary Street and Millbank estates. However, most activity in the great majority of British towns occurred after the turn of the century, making the council-house a twentieth- rather than a nineteenth-century phenomenon.

Conclusion

This chapter has attempted to signpost the growing involvement of local and national government in urban affairs in the nineteenth century. The presentation has been selective, but given the large amount of legislation relating to living conditions in towns and the changes which occurred in urban administration itself over the century, it was inevitable that the treatment would have to be partial. The selection of material has been guided by the wish to emphasize and illustrate two particular points, first, the progressive involvement of a variety of administrative bodies in matters which were thought to be badly served by the practices of the uncontrolled free market, and, second, the impact of such involvement on the urban built form. On the first point, it has been shown that local improvement commissions set up under Private Acts of Parliament were gradually superseded by reformed or new urban authorities using powers conferred on them by national legislation. The Public Health Acts of 1848 and 1875 were perhaps the two most significant pieces of legislation, but by the end of the century urban authorities were able to draw on a very considerable range of measures to deal with the living conditions in and the supply of utilities and services to towns. On the second point, the impact on the built form, this new legislation gradually made its mark on the urban environment. In individual towns, improvement schemes such as those described for Scottish towns, were put into practice, but more generally it was the by-laws which followed from the Public Health Acts which had the most pronounced universal influence in Britain. Returning again to Ashworth's observation on the 1875 Act, 'nothing else made so much difference to the physical appearance and condition of British towns' (Ashworth, 1954, 91).

13

Conclusion

This concluding chapter need only be brief. It is no more than a reiteration of the principles set out at the beginning in light of the content which has been developed. The fundamental basis from which all followed was the unprecedented increase in urban population. The locational pattern was initially virtually an haphazard one. It was dependent on where entrepreneurs could envisage maximizing profits and where such expectations were realized. 'The history of capitalism should be read as a series of locally – or regionally – specific investments, each moving through a cycle of innovation, adaptation, then crisis as capital moves on to areas of unused freedom' (Dodgshon, 1987, 324). Or as Harvey contends, 'the industrialization that ultimately subdued merchant capital was a new form of urbanism – a process in which Manchester, Leeds, and Birmingham were transformed from insignificant villages or minor towns, to industrial cities of great productive might' (Harvey, 1973, 259–60).

The consequence of the restless shifting of investment and of the escape from borough control was the complexity of an unintegrated series of regional developments creating a situation which demands local rather than national scale analysis. But, even so, from the beginning there were forces regulating and directing growth towards the formation of a national system of cities with a form of hierarchical structuring. Those forces ranged from a change in the nature of business towards more centralization, progressively moving from the single entrepreneur to the partnership and to the joint-stock company, to the organizational need for the marketing and the exporting of products, and the growing demand of the population for consumer goods of all sorts. There was, therefore, throughout the century a shift towards an integrated, national system of cities characterized at critical levels by diagnostic 'trait complexes' of standard stores, which slowly colonized the towns, and of services. It was a process which saw, for example, the high complexity of local banking reduced to the 'big five' which Smailes used in 1944 in his pioneering attempt to establish an urban hierarchy for England and Wales (Smailes, 1944). But as the date of Smailes's paper suggests, it was hardly accomplished by the end of the century. Nevertheless, although barely emergent by 1914 and subsequently greatly modified by developments after 1945, it still remains the basic substratum. However much was inherited from the mercantile era, there *was* an overturning and a transformation. The great cities of the nineteenth century, London excepted, were not the great cities of the seventeenth century, but they remain the great cities of the present day.

The massive increase of urban population, and the activities which both generated it and it generated, brought about greatly extended demands for land in the towns. At the forefront was the factory system creating large areas devoted to industry which, owing to restricted mobility and systems of casual labour, were shot through with working-class housing. At first much of that housing was of the most primitive quality. Slowly, however, through the century a combination of philanthropy, expressed as model towns and industrial villages, self-interest derived from a fear of disease and a possible deterioration in the quality of the work force, and of technical improvements in civil engineering, raised the quality. It is difficult to accept the description 'an age of mass consumption' as applicable to the century, but certainly demand for consumer goods and services increased to an extent that the shopping centre became a universal feature, characterized by national chain stores. Even in the smallest and most remote industrial villages the truck shop gave way to the little cluster of shops. In the larger towns the centre itself was transformed into a central business district by the proliferation of professional services, the growth of financial institutions and the great increase in the range of functions of local government.

All the changes which have just been noted inevitably extended occupations into a system of great complexity, but one in which each class, if that term can be used to subsume occupational groups, looked to associate itself with the level above and disassociate itself from the level below. Social distance because translated into effective physical distance on a regional scale within the city by the century's end. Anything approaching homogeneous social areas only existed at the extreme ends of the scale but, even so, as classes were separated by distance, or segregated, the necessity for and the common bases for co-operation which proximity had engendered were destroyed and class conflict became more an overt force in urban politics. But the progress of municipal reform ensured that the 'urban mob' became less of a significant element during the century so that conflict if more open was contained within legitimate bounds. The book edited by David Cannadine (1982) entitled *Patricians, power and politics in nineteenth-century towns* demonstrates admirably the way that the early power of the wealthy was modified during the century by the rise of elected urban governments. It also shows the way in which accommodation was reached between conflicting interests and the role of the patrician class was changed into one which was more honorary than real.

But demands for space of themselves meant very little without the necessary supply of land. That meant that owners and developers played a crucial role in the way that the city was formed during the century. In many cases they were able to mould the patterns of growth. In addition, the actual townscape was created by the builders who ranged from the modest erector of one or two houses to the great names of the Victorian era whose creations still dominate so many towns, even the political heart of the country at Westminister. And ever present and growing in significance was that political heart, intervening with legislation which was eventually to lead to the first Town Planning Act of 1909.

It was this highly complex set of controlling influences which was to determine the character of the towns of England and Wales prior to the First World War. But they operated in an infinite number of ways dependent upon local factors. Timing was one great variant. The first movement of the gentry out of the centre of Bristol, for example, can be identified in the early eighteenth century, it had barely started in Cardiff by the end of the nineteenth. There were similar processes at work, however, and both in land-use and residential location separation and segregation dominated so that in Chapter 9 it was possible to construct a model of the town towards the end of the nineteenth century showing how the common processes worked themselves out on the ground. But the homogeneous social areas which have been mentioned were the product of the changes brought about by the 1914–18 war, of much easier intra-urban transport which dispensed with horses and all those needed to look after them, and of a whole range of technical innovations in domestic appliances. Heating by gas and electricity, vacuum cleaners, washing machines, all meant that the panoply of servants and maids was no longer as necessary, even if they could be hired in the changed social milieu of the post-war world. But all these were the changes of a later period than the one which has been the subject of this book.

But if it is anachronistic to read modern social areas back into the nineteenth century, nevertheless the cities of that period exhibited contrast as one of their major features. That contrast was brought about by the great gap which arose between the powerful and the wealthy on the one hand and the poor and the destitute on the other. For the former, in spite of trade cycles, it was an age of exuberant confidence which was made manifest in the great town halls, the new churches and the elaborate gothic architecture of suburban houses; it was an age which created a Crystal Palace. The frenetic activity of mill and factory was echoed in the drive to express wealth in physical terms, especially by those to whom wealth was new. The term '*nouveau riche*' first appeared according to the Oxford English Dictionary in 1828. Even the 'respectable' working class stuffed their front parlours with mahogany furniture in emulation of the displays of their social superiors and as an indication of a status attained.

But the nineteenth century was also characterized by squalor and despair (Treble, 1979). This is how the correspondent of the *Morning Chronicle* described one of the worst areas of Merthyr Tydfil in 1850.

> There is a quarter of the town extending along a flat on the right bank of the Taff, from the lowest point of High Street towards Cyfarthfa – the proper name of which is Pont Storehouse; but like the unhappy and lawless people who inhabit it, the place has an alias, and is generally known by the name of 'China'. The houses are mere huts of stone – low, confined, ill-lighted, and unventilated; they are built without pretensions to regularity and form a maze of courts and tortuous lanes, hardly passable in many places, for house refuse, rubbish and filth. In some parts they are considerably below the level of the road and descent is by ladders. Such houses are called 'the cellars'. Here it is that in a congenial atmosphere, the crime, disease and penury of Merthyr are for the most part located. Thieves, prostitutes, vagrants, the idle, the reckless and the dissolute, here live in miserable companionship. This neighbourhood formed the main scene of our enquiries, and what I that day saw of misery, degradation and suffering, I shall remember to the end of my life. (*Morning Chronicle*, 'Labour and the Poor'. Letter IX, 29 April, 1850.)

It was, therefore, the best and the worst of times dependent upon social status. If there were truly homogeneous social areas then they were the working-class urban villages, often reliant on one industry which emphasized togetherness. That elusive thing called a feeling of community was built up, in many ways probably something akin to the strengthening and closing of relations which occur in time of disaster. It was also a product of limited mobility, although most evidence suggests that there was extensive migration at all scales.

All these characteristics are, of course, so effectively portrayed in the literature of the century. The Forsytes must surely epitomize the men of property and all the gloomy luxury of Victorian opulence. The Pooters depend for the gentle but affectionate humour they generate on the foibles of those who sought to exaggerate their social standing. The most prolific source is Charles Dickens, whose novels almost provide of themselves an urban geography of nineteenth-century England. *Bleak House* opens with an unforgettable picture of the urban street.

Smoke lowering down from chimney pots, making a soft black drizzle, with flakes of soot in it as big as full-grown snow-flakes – gone into mourning, one might imagine, for the death of the sun. Dogs, undistinguishable in mire. Horses, scarcely better; splashed to their very blinkers. Foot passengers, jostling one another's umbrellas, in a general infection of ill temper, and losing their foothold at street corners, where tens of thousands of other foot passengers have been slipping and sliding since the day broke (if ever the day broke), adding new deposits to the crust upon crust of mud, sticking at those points tenaciously to the pavement, and accumulating at compound interest. Fog everywhere. (Charles Dickens, 1852–53 *Bleak House*).

The novels people the streets with the characters which gave them life and movement. However, this reference to literary sources is not meant at this late stage to introduce an extended analysis of its relevance to urban studies, but rather to develop one final issue. This book has been firmly pitched within the field of urban geography. It is unashamedly and of necessity a limited field which does not claim to encompass all the aspects of urban life which are the domain of other specialisms. Nor does it have any pretence at adopting a humanistic approach, even if that were possible given the sorts of sources which are available. To a considerable degree the dissection which has taken place has been the dissection of a necessarily inert body. But the city was in reality alive and stopping of its daily rhythms for analytical purposes inevitably loses a great deal. In the book there has been little of what it was like to live in the city in the century under review. To most people at the time knowledge in any case was only partial. Although industrial areas are discussed there is nothing of the hideous noise and danger of factory and mill; there is nothing of the cramped misery and the perils of the mine; rivers and canals shown on the maps as black lines were stinking and polluted water courses –

It had a black canal in it, and a river that ran purple with ill-smelling dye. (Charles Dickens, *Hard Times*)

There is nothing of the dreariness of unremitting labour six days a week. There is nothing either of *The Best Circles* (Davidoff, 1973) or of the grandeur of high life in mock medieval follies; nothing of the reasons why one of the entrepreneurs responsible for the development of conditions described in the extract about 'China' at Merthyr Tydfil had carved on his massive tombstone the simple epitaph, 'God Forgive Me'.

But the background to all these conditions, in part being created by them, was the city constituted by shapes on the ground, moulded by and moulding society (Cannadine, 1982). There is at every point the temptation not only to widen discussion to the whole operation of urban society, but to the total economic and social system of which it was a part. And so one returns to the Introduction to the book and the purpose there outlined, the modest one of carrying out a dissection of urbanization and urbanism during the nineteenth century from the viewpoint of the urban geographer.

References

Adams, I. H. 1978: *The Making of Urban Scotland*. London: Croom Helm.

Adburgham, A. 1964: *Shops and Shopping, 1800–1914*. London: Allen & Unwin.

Alexander, D. 1970: *Retailing in England during the Industrial Revolution*. London: University of London, Athlone Press.

Allan, C. M. 1965: The genesis of British urban redevelopment with special reference to Glasgow. *Economic History Review*, 18 (Second series), 598–613.

Anderson, M. 1972: The study of family structure. In Wrigley, E. A. (ed.), *Nineteenth-Century Society*. Cambridge: Cambridge University Press.

Armstrong, W. A. 1972: The use of information about occupation. In Wrigley, E. A. (ed.), *Nineteenth-Century Society*. Cambridge: Cambridge University Press.

Ashworth, W. 1954: *The Genesis of Modern British Town Planning*. London: Routledge & Kegan Paul.

Aspinall, P. J. (n.d.): *The evolution of urban tenure systems in nineteenth century cities*. Research memorandum no. 63. Centre for Urban and Regional Studies, University of Birmingham.

——1978: *Building Applications and the Building Industry in Nineteenth-Century Towns: The Scope for Statistical Analysis*. Research memorandum no. 68, Centre for Urban and Regional Studies, University of Birmingham.

Aspinall, P. J. and Whitehand, J. W. R. 1980: Building plans: a major source for urban studies. *Area*, 12, 199–203.

Dr Ballard's Report to the Local Government Board on the sanitary condition of the Municipal Borough of Dudley. 1874.

Dr Ballard's Report to the Local Government Board on the sanitary condition of the Municipal Borough of Wolverhampton. 1874.

Barber, B. J. 1980: Aspects of municipal government, 1835–1914. In Fraser, D. (ed.), *A History of Modern Leeds*. Manchester: Manchester University Press, 301–26.

Barker, D. 1978: A conceptual approach to the description and analysis of an historical urban system. *Regional Studies*, 12, 1–10.

——1980: Structural change in hierarchic spatial systems in south-west England between 1861 and 1911. *Geog. Annaler*, 623, 1–9.

Bell, C. and R. 1972: *City Fathers: The Early History of Town Planning in Britain*. Harmondsworth: Penguin Books.

Beresford, M. W. 1971: The back-to-back house in Leeds, 1787–1937. In Chapman, S. D. (ed.), *The History of Working-Class Housing*. Newton Abbot: David & Charles.

Black, I. 1989: Geography, political economy and the circulation of finance capital in early industrial England. *Journ. Hist. Geog.*, 15(4), 368–84.

Bowden, M. J. 1975: Growth of the Central Districts in Large Cities. In Schnore, L. F. and Lampard, E. E. (eds.), *The New Urban History*, 75–109. Princeton, NJ: Princeton University Press.

Braudel, F. 1982: *Civilization and Capitalism 15th–18th Century*. Vol. 2. *The Wheels of Commerce*. Trans. Sian Reynolds, London: Collins.
——1984: *Civilization and Capitalism 15th–18th Century*. Vol. 3. *The Perspective of the World*. London: Collins.
Briggs, A. 1963: *Victorian Cities*. London: Odhams Press.
——1968: Public health: the sanitary idea. *New Society*, 281, 229–31.
Broaderwick, R. F. 1981: *An Investigation into the Location of Institution Land Uses in Birmingham*. University of Birmingham. Ph.D. thesis, unpub.
Buck, N. H. 1981: The analysis of state intervention in nineteenth-century cities. In Dear, M. and Scott, A. J. (eds.), *Urbanisation and Urban Planning in Capitalist Society*, 501–34, London: Methuen.
Buckingham, J. S. 1849: *National Evils and Practical Remedies*. London: Jackson.
Bucklin, L. P. 1972: *Competition and Evolution in the Distributive Trades*. Englewood Cliffs: Prentice-Hall.
Burke, G. 1971: *Towns in the Making*. London: Edward Arnold.
——1976: *Townscapes*. Harmondsworth: Pelican Books.
Burnett, J. 1980: *A Social History of Housing 1815–1970*. London: Methuen.
Butterworth, J. 1822: *The antiquities of the town and a complete history of the trade of Manchester: with a description of Manchester and Salford*. Manchester, Printed for the author.
Bythell, D. 1978: *The Sweated Trades. Outwork in Nineteenth-Century Britain*. London: Batsford Academic.
Cannadine, D. 1977: Victorian cities: how different? *Social History*, 2, 457–82.
——1980: *Lords and Landlords: The Aristocracy and the Towns 1774–1967*. Leicester: Leicester University Press.
——1982a: *Patricians, Power and Politics in Nineteenth Century Towns*. Leicester; Leicester University Press.
——1982b: Residential differentiation in nineteenth-century towns: from shapes on the ground to shapes in society. In Johnson, J. H. and Pooley, C. G. (eds.), *The Structure of Nineteenth Century Cities*. London: Croom Helm.
Cannadine, D. and Reeder, D. (eds.) 1982: *Exploring the urban past. Essays in Urban History by H. J. Dyos*. Cambridge: Cambridge University Press.
Caroe, L. 1968: A multivariate grouping scheme: association analysis of East Anglian towns. In Bowen, E. G., Carter, H. and Taylor, J. A. (eds.), *Geography at Aberystwyth*, 252–369. Cardiff: University of Wales Press.
Carter, H. 1956: The urban hierarchy and historical geography. *Geog. Studies*, 3, 85–101.
——1965: *The Towns of Wales. A Study in Urban Geography*. Cardiff: University of Wales Press.
——1969: *The Growth of the Welsh City System*. Cardiff: University of Wales Press.
——1980: Transformations in the spatial structure of Welsh towns in the nineteenth century. *Transactions of the Honourable Society of Cymmrodorion*, 175–200.
——1981: *The Study of Urban Geography*, 3rd edition. London: Edward Arnold.
——1983: *An Introduction to Urban Historical Geography*. London: Edward Arnold.
Carter, H. and Wheatley, S. 1982: *Merthyr Tydfil in 1851*. Cardiff: University of Wales Press.
Carter, H. and Lewis, C. R. 1983: *Processes and Patterns in Nineteenth Century Cities*. Unit 15 in Open University course D301, Historical sources and the social scientist. Milton Keynes: Open University Press.
Catalogue of British Parliamentary Papers in the Irish University Press 1000 – Volume Series and Area Studies Series, 1801–1900 1977: Dublin: Irish University Press.
Chalklin, C. W. 1974: *The Provincial Towns of Georgian England*. London: Edward Arnold.
Chambers, J. D. 1952: *A Century of Nottingham History, 1851–1951*. Nottingham: University of Nottingham.
Chapman, S. D. 1974: *Jesse Boot of Boots the Chemist*. London: Hodder & Stoughton.
Chapman, S. D. (ed.) 1971: *A History of Working-Class Housing*. Newton Abbot: David & Charles.

Chappell, E. L. 1946: *Cardiff's Civic Centre: A Historical Guide*. Cardiff: Priory Press.

Cleary, E. J. 1965: *The Building Society Movement*. London: Elek Books.

Conzen, M. R. G. 1960: Alnwick, Northumberland: a study in town-plan analysis. *Transactions of the Institute of British Geographers*, 27, 1–122.

Cottrell, P. L. 1986: Banking and finance. In Langton, J. and Morris, R. J. (eds.), *Atlas of Industrializing Britain*. London: Methuen, 144–55.

Court, W. H. B. 1954: *A Concise Economic History of Britain from 1750 to Recent Times*. Cambridge: Cambridge University Press.

Cowlard, K. A. 1979: The identification of social (class) areas and their place in nineteenth-century urban development. *Transactions of the Institute of British Geographers* (New Series), 4, 239–57.

Crossick, G. 1978: *An Artisan Elite in Victorian Society*. London: Croom Helm.

Cunningham, C. 1981: *Victorian and Edwardian Town Halls*. London: Routledge & Kegan Paul.

Daunton, M. J. 1977: *Coal Metropolis: Cardiff, 1870–1914*. Leicester: Leicester University Press.

——1983: *House and Home: Working Class Housing, 1850–1914*. London: Edward Arnold.

Davidoff, L. 1973: *The Best Circles: Society, Etiquette and the Season*. London: Croom Helm.

Davies, J. 1981: *Cardiff and the Marquesses of Bute*. Cardiff: University of Wales Press.

Davies, J. R. 1976: 'Bryn Wyndham' village, upper Rhondda Fawr. *Morgannwg*, 20, 53–65.

Davies, W. K. D. 1970: Toward an integrated study of central places. In Carter, H. and Davies, W. K. D. (eds.), *Urban Essays. Studies in the Geography of Wales*, 193–227. London: Longman.

Davies, W. K. D., Giggs, J. A. and Herbert, D. T. 1968: Directories, rate books and the commercial structure of towns. *Geography*, 53, 41–54.

Davis, D. 1966: *A History of Shopping*. London and Toronto: Routledge & Kegan Paul.

Davis, H. and Scase, R. 1985: *Western Capitalism and State Socialism*. Oxford: Blackwell.

Dear, M. and Scott, A. J. (eds.) 1981: *Urbanization and Urban Planning in Capitalist Society*. London: Methuen.

Dennis, R. 1984: *English Industrial Cities of the Nineteenth Century: A Social Geography*. Cambridge: Cambridge University Press.

Dennis, R. and Clout, H. 1980: *A Social Geography of England and Wales*. Oxford: Pergamon.

de Vries, J. 1981: Patterns of urbanization in pre-industrial Europe. In Schmal, H. (ed.), *Patterns of European Urbanization since 1500*, 77–110. London: Croom Helm.

——1984: *European Urbanization 1500–1800*. Cambridge, Mass.: Harvard University Press.

Dillon, T. 1974–79: The Irish in Leeds, 1851–1861. *Publications of the Thoresby Society*, 54, 1–28.

Dodgshon, R. A. 1987: *The European Past. Social Evolution and Spatial Order*. London and Basingstoke: Macmillan.

Doughty, M. (ed.) 1986: *Building the Industrial City*. Leicester: Leicester University Press.

Dyos, H. J. 1955: Railways and housing in Victorian London. *Journal of Transport History* 2, 11–21 and 90–100.

——1957: Some social costs of railway building in London. *Journal of Transport History*, 3, 23–30.

——1961: *Victorian Suburb: A Study of the Growth of Camberwell*. Leicester: Leicester University Press.

——1968: The speculative builders and developers of Victorian London. *Victorian Studies*, 11 (supplement), 641–90.

Dyos, H. J. and Wolff, M. (eds.) 1982: *The Victorian City. Images and Realities*, 2 vols. London: Routledge & Kegan Paul.

Edwards, A. M. 1981: *The Design of Suburbia*. London: Pembridge Press.

Edwards, K. C. (ed.) 1966: *Nottingham and its Region*. Nottingham: British Association.

Engles, F. 1969: *The Condition of the Working Class in England*. London: Panther. Originally published in German in 1845.

Errazurez, A. 1946: Some types of housing in Liverpool. *Town Planning Review*, 19, 57–68.

Falkus, M. 1977: The development of municipal trading in the nineteenth century. *Business History*, 19, 134–61.

Farrant, S. Fossey, K. and Peasgood, A. 1981: *The Growth of Brighton and Hove, 1840–1939*. Brighton: University of Sussex.

First Report of the Commissioners on the State of Large Towns and Populous Districts, 1844. 1970: Irish University Press Series of British Parliamentary Papers. Health, General, Vol. 5. Shannon: Irish University Press.

Flinn, M. W. (ed.) 1965: *Report on the Sanitary Condition of the Labouring Population of Great Britain by Edwin Chadwick, 1842*. Edinburgh: Edinburgh University Press.

Forster, C. A. 1972: *Court Housing in Kingston upon Hull*. Occasional Papers in Geography No. 19, University of Hull.

Fox, R. C. 1979: The morphological, social and functional districts of Stirling, 1798–1881. *Transactions of the Institute of British Geographers* (New Series), 4, 153–67.

Fraser, D. and Sutcliffe, A. (eds.) 1983: *The Pursuit of Urban History*. London: Edward Arnold.

Fraser, W. H. 1981: *The Coming of the Mass Market 1850–1914*. London: Macmillan.

Fried, M. H. 1967: *The Evolution of Political Society*. New York: Random House.

Friedlander, D. 1970: The spread of urbanization in England and Wales 1851–1951. *Population Studies*, 24(3), 423–43.

Frost, P. M. 1973: *The Growth and Localization of Rural Industry in South Staffordshire, 1500–1720*. University of Birmingham, Unpub. PhD thesis.

Gauldie, E. 1974: *Cruel Habitations*. London: Allen & Unwin.

Geddes, P. 1949: *Cities in Evolution*. London: Williams & Norgate. New and revised edition of the original 1915 publication.

Gordon, G. 1979: The status areas of early to mid-Victorian Edinburgh. *Transactions of the Institute of British Geographers* (New Series), 4, 168–91.

Gray, D. 1953: *Nottingham: Settlement to City*. Nottingham: Nottingham Co-operative Society Ltd.

Gregory, D. 1982: *Regional Transformation and Industrial Revolution*. London and Basingstoke: Macmillan.

Gregory, D. 1988a: The Production of regions in England's Industrial Revolution: a reply. *Journal of Historical Geography*, 14, 50–58.

Gregory, D. 1988b: The Production of regions in England's Industrial Revolution. *Journal of Historical Geography*, 14, 174–176.

Hall, P. 1962: *The Industries of London since 1861*. London: Hutchinson.

Halstead, K. A. 1982: *The Economic Factors in the Development of Urban Fabric of London's Docklands, 1796–1909*. City of London Polytechnic, PhD thesis, unpub.

Harley, J. B. 1972: *Maps for the Local Historian: A Guide to the British Sources*. London: National Council of Social Service.

Harley, J. B. and Phillips, C. W. 1964: *The Historian's Guide to Ordnance Survey Maps*. London: National Council of Social Service.

Harvey, D. 1973: *Social Justice and the City*. London: Edward Arnold.

Hirsch, J. 1981: The apparatus of the state. The reproduction of capital and urban conflicts. In Dear, M. and Scott, A. J. (eds.), *Urbanization and Urban Planning in Capitalist Society*, 583–608. London: Methuen.

Hobhouse, H. 1971: *Thomas Cubitt, Master Builder*. New York: Universe Books.

Holmes, R. S. 1973: Ownership and migration from a study of rate books. *Area*, 5, 242–51.

Homan, R. 1983: The early development of the building society movement in the Cheltenham region. *Bristol and Gloucestershire Archaeological Society Transactions*, 101, 161–70.

Homan, R. and Rowley, G. 1979: The location of institutions during the process of urban growth, a case study of churches and chapels in nineteenth century Sheffield. *The East Midland Geographer*, 7(4), 137–52.

Hughes, J. 1970: *Industrialization and Economic History: Theses and Conjectures*. New York: McGraw Hill.

Jeffreys, J. B. 1954: *Retail Trading in Britain, 1850–1950*. Cambridge: Cambridge University Press.

Jones, E. 1960: *A Social Geography of Belfast*. London: Oxford University Press.

Jones, P. N. 1969: *Colliery Settlement in the South Wales Coalfield 1850 to 1926*. University of Hull, Occasional Papers in Geography No. 14.

Kaye, B. 1960: *The Development of the Architectural Profession in Britain*. London, Allen & Unwin.

Keith-Lucas, B. 1953–54: Some influences affecting the development of sanitary legislation in England. *Economic History Review*, 2nd series, 6, 290–6.

——1980: *The Unreformed Local Government System*. London: Croom Helm.

Kellett, J. R. 1969: *The Impact of Railways on Victorian Cities*. London: Routledge & Kegan Paul.

Keyfitz, N. 1980: Do cities grow by natural increase or by migration? *Geog. Analysis*, 12, 143–56.

Keyfitz, N. and Dimiter, P. 1981: Migration and natural increase in the growth of cities. *Geog. Analysis*, 13, 288–99.

Killick, J. R. and Thomas, W. A. 1970: The Provincial Stock Exchanges. *Economic History Review*, 2nd series, 23, 96–111.

Krausz, E. 1964: *Leeds Jewry: Its History and Social Structure*. Cambridge: Heffer.

Langton, J. 1972: Coal output in South-West Lancashire 1590–1799. *Economic History Review*, Second Series 25, 28–54.

——1984: The industrial revolution and the regional geography of England. *Transactions of the Institute of British Geographers* (New Series), 9, 145–167.

——1988: The production of regions in England's Industrial Revolution: a response. *Journal of Historical Geography*, 14, 170–73.

Langton, J. and Morris, R. J. (eds) 1986: *Atlas of Industrializing Britain*. London: Methuen.

Large, P. 1985: Urban growth and agricultural change in the West Midlands during the seventeenth and eighteenth centuries. In Clark, P. (ed.), *The Transformation of English Provincial Towns*, 169–89. London: Hutchinson.

Law, C. M. 1967: The growth of urban population in England and Wales, 1801–1911. *Transactions of the Institute of British Geographers*, 41, 125–43.

Lawton, R. 1968: Population changes in England and Wales in the later nineteenth century: an analysis of trends by registration districts. *Transactions of the Institute of British Geographers*, 44, 55–74.

——(ed.) 1978: *The Census and Social Structure*. London: Cass.

——1983: Urbanisation and population change in nineteenth century England. In Patten, J. (ed.), *The Expanding City. Essays in Honour of Professor Jean Gottmann*. London: Academic Press.

——1986: Population. In Langton, J. and Morris, R. J. (eds), *Atlas of Industrializing Britain*, 10–29. London: Methuen.

Lawton, R. and Pooley, C. G. 1974: David Brindley's Liverpool: an aspect of urban society in the 1880s. *Transactions of the Historic Society of Lancashire and Cheshire*, 125, 149–68.

——1976: *The Social Geography of Merseyside in the Nineteenth Century*. Final report to the Social Science Research Council (Britain Lending Library).

Lewis, C. R. 1970: The central place pattern of Mid-Wales and the Middle Welsh Borderland. In Carter, H. and Davies, W. K. D. (eds), *Urban Essays*, 228–68. London: Longman.

——1975: The analysis of changes in urban status: a case study in Mid-Wales and the middle Welsh borderland. *Transactions of the Institute of British Geographers*, 64, 49–65.

——1975: Trade directories – a data source in urban analysis. *The National Library of Wales Journal*, 19, 181–93.

——1979: A stage in the development of the industrial town: a case study of Cardiff, 1845–75. *Transactions of the Institute of British Geographers* (New series), 4, 129–52.

——1980: The Irish in Cardiff in the mid-nineteenth century. *Cambria*, 7, 13–41.

——1985a: Locational patterns of high status groups in an emerging industrial town: Newport (Gwent), 1850–1880. *Cambria*, 12, 131–47.

——1985b: Housing areas in the industrial town: a case study of Newport, Gwent, 1850–1880. *National Library of Wales Journal*, 24, 118–45.

——1986: Changes in the central place system of Mid-Wales: a consideration of data and trends. In Grant, E. (ed.), *Central Places, Archaeology and History*. Sheffield: Department of Archaeology, University of Sheffield.

Lipman, V. D. 1949: *Local Government Areas 1834–1945*. Oxford: Blackwell.

——1968: The rise of Jewish suburbia. *Transactions of the Jewish Historical Society of England*, 21, 78–103.

Lowe, J. B. 1977: *Welsh Industrial Workers Housing 1775–1875*. Cardiff: National Museum of Wales.

MacKeith, M. 1985: *Shopping Arcades*. London: Mansell.

McKendrick, M., Brewer, J. and Plumb, J. H. 1982: *The Birth of a Consumer Society. The Commercialization of Eighteenth Century England*. London: Europa Publications.

Mills, D. 1982: *A Guide to Census Enumerators' Books*. Milton Keynes: Open University Press.

Mills, D. and Pearce, C. 1989: *People and Places in the Victorian Census. A Review and Bibliography of Publications Based Substantially on the Manuscript Census Books, 1841–1911*. Institute of British Geographers Historical Geography Research Series, No. 23.

Mokyr, J. 1987: Has the Industrial Revolution been crowded out? *Explorations in Economic History*. Cambridge: Cambridge University Press.

Morgan, E. V. and Thomas, W. A. 1969: *The Stock Exchange: Its History and Function* 2nd Ed. London: Paul Elek.

Morgan, P. H. 1985: *Building use and social change in the inner city: a case study of Cardiff, 1850–1950*. Unpub. PhD thesis, University of Wales.

Morris, R. J. 1986: Urbanization. In Langton, J. and Morris, R. J. (eds.), *Atlas of Industrializing Britain*. 164–79. London: Methuen.

Mortimore, M. J. 1969: Landownership and urban growth in Bradford and its environs in the West Riding conurbation, 1850–1950. *Transactions of the Institute of British Geographers*, 46, 105–19.

Mounfield, P. R. 1967: *Footwear Industry of the East Midlands*. mss in Library, U.C.W. Aberystwyth.

Mui, H. C. and Mui, L. H. 1989: *Shops and Shopping in Eighteenth Century England*. London and Montreal: McGill-Queens University Press.

Muir, R. 1910: Liverpool, an analysis of the geographical distribution of civic functions. *Town Planning Review*, 1,304.

Muthesius, S. 1982: *The English Terraced House*. New Haven: Yale University Press.

Olsen, D. J. 1982: *Town Planning in London*. New Haven: Yale University Press.

D. Owen and Co's (Wrights) Cardiff Directory 1890. Cardiff: Western Mail.

Paine, H. J. 1856: *Third Annual Report on the Sanitary Condition of Cardiff (for the year 1855) by H.J. Paine, Officer of Health to the Local Board of Health*. Cardiff.

Payne, P. L. 1988. *British Entrepreneurship in the Nineteenth Century*. 2nd edition, London: MacMillan.

Pollard, S. 1981: *Peaceful Conquest: the Industrialization of Europe 1760–1970*. Oxford: Oxford University Press.

Polyani, K. 1968: *Primitive, Archaic and Modern Economies: Essays of K. Polyani*, ed. Dalton, G. Boston, Mass.: Beacon Press.

Pooley, C. G. 1977: The residential segregation of migrant communities in mid-Victorian Liverpool. *Transactions of the Institute of British Geographers* (New series), 2, 364–82.

——1979: Residential mobility in the Victorian city. *Transactions of the Institute of British Geographers* (New series), 4, 258–277.

——1983: Welsh migration to England in the mid-nineteenth century. *Journal of Historical Geography*, 9, 287–306.

Powell, C. G. 1982: *An Economic History of the British Building Industry, 1815–1979*. London: Methuen.

Pred, A. 1977: *City Systems in Advanced Economies*. London: Hutchinson.

Price, S. J. 1958: *Building Societies: Their Origin and History*. London: Franey & Co.

Rammell, T. W. 1850: *Report to the General Board of Health on a Preliminary Inquiry into the Sewerage, Drainage, and Supply of Water, and the Sanitary Condition of the Inhabitants of the Town of Cardiff*. London: HMSO.

Ravenstein, E. G. 1885: The Laws of Migration. *Journal of the Stat. Soc. of London*, 48, 167–227.

Redlich, J. and Hirst, F. W. 1970: *The history of local government in England*. London: MacMillan.

Rees, J. F. 1957: *The Story of Milford*. Cardiff: University of Wales Press.

Report from the Select Committee on Town Holdings, 1887 1969: Irish University Press Series of British Parliamentary Papers. Urban Areas, Planning, 5. Shannon: Irish University Press.

Report from the Select Committee on Town Holdings, 1889 1969: Irish University Press Series of British Parliamentary Papers. Urban Areas, Planning, 7. Shannon: Irish University Press.

Richardson, C. 1971: The Irish in Victorian Bradford. *Bradford Antiquary*, 11, 294–316.

Roberts, R. 1982: Leasehold estates and municipal enterprise: landowners, local government and the development of Bournemouth, c. 1850 to 1914. In Cannadine, D. (ed.), *Patricians, Power and Politics in Nineteenth-Century Towns*, 176–218. Leicester: Leicester University Press.

Robson, B. T. 1966:. An ecological analysis of the evolution of residential areas in Sunderland. *Urban Studies*, 3, 120–42.

——1973: *Urban Growth: An Approach*. London: Methuen.

——1981: The impact of functional differentiation within systems of industrial cities. In Schmal, H. (ed.), *Patterns of European Urbanization since 1500*, 111–30. London: Croom Helm.

Rodger, R. G. 1979: Speculative builders and the structure of the Scottish building industry, 1860–1914. *Business History*, 21, 226–46.

——1983: The invisible hand: market forces, housing and the urban form in Victorian cities. In Fraser, D. and Sutcliffe, A. (eds.), *The Pursuit of Urban History*. London: Edward Arnold.

Rodgers, H. B. 1961–2: The suburban growth of Victorian Manchester. *Transactions of the Manchester Geographical Society*, 1–12.

Rostow, W. W. 1960: *The Stages of Economic Growth*. Cambridge: Cambridge University Press.

Rowley, G. 1984: *British Fire Insurance Plans*. Old Hatfield, Herts.: Goad Ltd.

Royal Commission on the Housing of the Working Classes, 1885. 1970: Irish University Press Series of British Parliamentary Papers. Urban Areas, Housing, 2. Shannon: Irish University Press.

Royle, S. A. 1977: Social stratification from early census returns: a new approach. *Area*, 9, 215–19.

Rubinstein, D. 1974: *Victorian Homes*. Newton Abbot: David & Charles.

Scola, R. 1975: Food markets and shops in Manchester 1770–1870. *Journal of Historical Geography*, 1, 153–67.

Scottish Land 1914: *The Report of the Scottish Land Enquiry Committee*. London: Hodder & Stoughton.

Searle, M. (n.d.): *Turnpikes and Toll-bars*, 2 vols. London: Hutchinson.

Schmal, H. 1981: *Patterns of European Urbanization since 1500*. London: Croom Helm.

Shaw, G. 1978: Patterns and processes in the geography of retail change with special reference to Kingston-upon-Hull. University of Hull, *Occasional Papers in Geography*, 4.

——1982: *British Directories as Sources in Historical Geography*. Historical Geography Research Series, No. 8. Norwich: Geo Abstracts.

——1988: Recent research on the commercial structure of nineteenth century British cities. In Denecke, D. and Shaw, G. (eds.), *Urban Historical Geography. Recent Progress in Britain and Germany*. 236–52. Cambridge: Cambridge University Press.

——1989: Industrialization, urban growth and the city economy. In Lawton, R. (ed.), *The Rise and Fall of Great Cities*. London and New York: Bellhaven Press.

Shaw, G. and Tipper, A. 1989: *British Directories. A Bibliography and Guide to Directories Published in England and Wales (1850–1950) and Scotland (1773–1950)*. Leicester: Leicester University Press.

Shaw, G. and Wild, M. T. 1979: Retail patterns in the Victorian city. *Transactions of the Institute of British Geographers* (New series), 4, 278–91.

Shaw, M. 1979: Reconciling social and physical space: Wolverhampton 1871. *Transactions of the Institute of British Geographers* (New Series), 4, 192–213.

Slater, T. R. 1977: Landscape parks and the form of small towns in Great Britain. *Transactions of the Institute of British Geographers* (New Series), 2, 314–31.

Smailes, A. E. 1944: The urban hierarchy of England and Wales. *Geography*, 29, 41–7.

Smith, C. T. 1951: The movement of population in England and Wales in 1851 and 1861. *The Geographical Journal*, 117(2), 200–10.

Smith, P. J. 1980: Planning as environmental improvement: slum clearance in Victorian Edinburgh. In Sutcliffe, A. (ed.), *The Rise of Modern Urban Planning*. London: Mansell.

Smith, R. 1974: Multi-dwelling building in Scotland 1750–1970: a case study based on housing in the Clyde Valley. In Sutcliffe, A. (ed.), *Multi-Storey Living*, 207–43. London: Croom Helm.

Spring, D. 1971: English landowners and nineteenth-century industrialism. In Ward, J. T. and Wilson, R. G. (eds.), *Land and Industry*. Newton Abbot: David & Charles.

Springett, R. J. 1979: *The Mechanics of Urban Land Development in Huddersfield, 1770–1911*. University of Leeds, PhD thesis, unpub.

Springett, R. J. 1982: Landowners and urban development: the Ramsden estate and nineteenth-century Huddersfield. *Journal of Historical Geography*, 8, 129–44.

Stapleton, H. E. C., Pace, G. G. and Day, J. E. 1975: *A Skilful Master Builder*. York: Ebor Press.

Stephens, W. B. 1981: *Sources for English Local History*. Cambridge: Cambridge University Press.

Sutcliffe, A. (ed.) 1980: *The Rise of Modern Urban Planning*. London: Mansell.

——1984: *Metropolis 1890–1940*. London: Mansell.

Tarn, J. N. 1973: *Five Per Cent Philanthropy*. Cambridge: Cambridge University Press.

——1980: Housing reform and the emergence of town planning in Britain before 1914. In Sutcliffe, A. (ed.), *The Rise of Modern Urban Planning*. London: Mansell.

Thernstrom, S. and Sennett, R. (eds.) 1969: *Nineteenth-Century Cities: Essays in the New Urban History*. New Haven, Conn.: Princeton University Press.

Thompson, F. M. L. 1968: *Chartered Surveyors, the Growth of a Profession*. London: Routledge & Kegan Paul.

Treble, J. H. 1979: *Urban Poverty in Britain 1830–1914*. London: Batsford Academic.

Vance, J. E. 1971: Land assignment in pre-capitalist, capitalist and post-capitalist cities. *Econ. Geog.*, 47, 101–20.

Varley, R. 1968: *Land Use Analysis in the City Centre with Special Reference to Manchester*. University of Wales, unpub. MA thesis.

Waller, P. J. 1983: *Town, City and Nation. England 1850–1914*. Oxford: Oxford University Press.

Wallerstein, I. 1974: *The Modern World System*. Vol. 1 *Capitalist Agriculture and the Origins of the European World-Economy in the Sixteenth Century*. New York and London: Academic Press.

——1979: *The Capitalist World Economy*. Cambridge: Cambridge University Press.

——1980: *The Modern World System. Vol. 2 Mercantilism and the Consolidation of the European World-Economy 1600–1750*. New York and London: Academic Press.

Ward, D. 1962: The pre-urban cadaster and the urban pattern of Leeds. *Annals of the Association of American Geographers*, 52, 150–66.

——1966: The industrial revolution and the emergence of Boston's Central Business District. *Economic Geography*, 42, 152–71.

——1971: *Cities and Immigrants. A Geography of Change in Nineteenth Century America*. New York: Oxford University Press.

——1975: Victorian cities: how modern? *Journal of Historical Geography*, 1, 135–51.

——1980: Environs and neighbours in the 'Two Nations': residential differentiation in mid-nineteenth-century Leeds. *Journal of Historical Geography*, 6, 133–62.

Warnes, A. M. 1973: Residential patterns in an emerging industrial town. In Clark, B. D. and Gleave, M. B. (eds.), *Social Patterns in Cities*. 169–189, London: Institute of British Geographers.

Webb, S. and B. 1922: *English Local Government: Statutory Authorities for Special Purposes*. London: Longmans, Green & Co.

Weber, A. F. 1899: *The Growth of Cities in the Nineteenth Century. A Study in Statistics*. New York: MacMillan. Columbia University, Studies in History, Economics and Public Land, Vol. 11. Reprinted 1967, New York, Cornell University Press.

Wheatley, S. E. 1983: *The Social and Residential Areas of Merthyr Tydfil in the Mid-Nineteenth Century*. University of Wales. PhD thesis, unpub.

Whitehand, J. W. R. 1967: Fringe belts: a neglected aspect of urban geography. *Transactions Institute of British Geographers*, 41, 223–33.

——1972: Urban-rent theory, time series and morphogenesis, an example of eclecticism in geographical research. *Area*, 4, 215–22.

——1987: The changing face of cities. A study of development cycles and urban form. *Institute of British Geographers Special Publication Series* No. 21. Oxford: Blackwell.

Williams, B. 1985: *The Making of Manchester Jewry, 1740–1875*. Manchester: Manchester University Press.

Wirth, L. 1938: Urbanization as a way of life. *American Journal of Sociology*, 44, 1–24.

Wise, M. J. 1949: On the evolution of the jewellery and gun quarters in Birmingham. *Transactions of the Institute of British Geographers* 15, 57–72.

Wise, M. J. and Thorpe, P. U. N. 1950: The growth of Birmingham 1800–1850. In Kinvig, R. H. *et al.* (eds.), *Birmingham and its Regional Setting. A Scientific Survey*. Birmingham: British Association for the Advancement of Science.

Wrigley, E. A. 1969: *Population and History*, London: Weidenfeld and Nicolson.

Wrigley, E. A. 1978: A simple model of London's importance in changing English society and economy, 1650–1750. In Adams P. A. and Wrigley, E. A. (eds.), *Towns in Societies: Essays in Economic History and Historical Sociology*. Cambridge: Cambridge University Press.

Wrigley, E. A. and Schofield, R. S. 1989: *The Population History of England 1541–1871*, 2nd edn. Cambridge: Cambridge University Press.

Yeadell, M. H. 1986: Building societies in the West Riding of Yorkshire and their contribution to housing provision in the nineteenth century. In Doughty, M. (ed.), *Building the Industrial City*. Leicester: Leicester University Press.

Index

88326   307.760942
                         C24

		DATE DUE	

Alliance Theological Seminary
Nyack, N.Y. 10960